U.S. Department
of Transportation

Research and Innovative Technology
Administration

**Volpe National Transportation
Systems Center**

CRASH SAFETY ASSURANCE STRATEGIES FOR FUTURE PLASTIC AND COMPOSITE INTENSIVE VEHICLES (PCIVs)

Final Report	June 2010
DOT-VNTSC-NHTSA-10-01	

Notice

This document is disseminated under the sponsorship of the Department of Transportation in the interest of information exchange. The United States Government assumes no liability for its contents or use thereof.

Volpe National Transportation Systems Center Cambridge, MA 02142	This document contains preliminary information subject to change. It is intended for technical communication within the Volpe Center and with NHTSA sponsors, with select distribution controlled by the sponsor.

298-102 REPORT DOCUMENTATION PAGE	
Public reporting burden for this collection of information is estimated to average 1 hour per response, including the time for reviewing instructions, searching existing data sources, gathering and maintaining the data needed, and completing and reviewing the collection of information Send comments regarding this burden estimate or any other aspect of this collection of information, including suggestions for reducing this burden, to Washington Headquarters Services, Directorate for Information Operations and Reports, 1215 Jefferson Davis Highway, Suite 1204, Arlington, VA 22202-4302, and to the Office of Management and Budget, Paperwork Reduction Project (0704-0188), Washington, DC 20503	
1 AGENCY USE ONLY (Leave blank)	2 REPORT DATE

4 TITLE AND SUBTITLE
Crash Safety Assurance Strategies For Future Plastic and Composite Intensive Vehicles (PCIVs)

6 AUTHOR(S)
Graham Barnes BEng CEng MIMechE
Ian Coles B.Tech (Hons)
Richard Roberts BSc CEng MIMechE
Daniel O. Adams, Ph.D.
David M. Garner, Jr., Ph.D.

7 PERFORMING ORGANIZATION NAME(S) AND ADDRESS(ES)
Engenuity Limited
The Old Hospital, Ardingly Road
Cuckfield, West Sussex RH17 5HF Great Britain
University of Utah
Department of Mechanical Engineering
50 S. Central Campus Drive
Salt Lake City, UT 84112 USA

9 SPONSORING/MONITORING AGENCY NAME(S) AND ADDRESS(ES)
U.S. Department of Transportation
National Highway Traffic Safety Administration

11 SUPPLEMENTARY NOTES

12a DISTRIBUTION/AVAILABILITY STATEMENT
This document is available to the public through the National Technical Information Service, Springfield, Virginia 22161.

13 ABSTRACT (Maximum 200 words)
This report identifies outstanding safety issues and research needs for Plastics and Composite Intensive Vehicles (PCIV) to facilitate their safe deployment by 2020. A PCIV definition is proposed, which ensures that the weight and efficiency objectives are prerequisite. Potential safety benefits of automotive plastics and composites are reviewed, and safety specifications associated with each level of the Building Block approach are presented. Lessons learned from the racing industry and from limited production, high-performance supercars with extensive use of composite materials are summarized. Changes and additions to test and evaluation procedures due to PCIVs are discussed, with a focus on ensuring their compliance with Federal Motor Vehicle Safety Standards (FMVSS). Progress is summarized and research recommendations proposed in three topic areas pertinent to crashworthiness of PCIVs: material databases, crashworthiness test method development, and crash modeling.

14 SUBJECT TERMS
Automotive crash safety; plastics and composites intensive vehicles (PCIV); light-weighting advanced materials; crash safety standards

17 SECURITY CLASSIFICATION OF REPORT	18 SECURITY CLASSIFICATION OF THIS PAGE	19 SECURITY CLASSIFICATION OF ABSTRACT
Unclassified	Unclassified	Unclassified

Notice

This document is disseminated under the sponsorship of the Department of Transportation in the interest of information exchange. The United States Government assumes no liability for its contents or use thereof.

PREFACE AND ACKNOWLEDGEMENTS

This report addresses outstanding safety issues and research needs for Plastics and Composite Intensive Vehicles (PCIVs) to facilitate their safe deployment by 2020. PCIVs have the potential to revolutionize the automotive sector; however, the use of plastics and composite materials in automotive structures requires an in-depth knowledge of their unique performance characteristics in the crash and safety environment. Included in this report is a proposed definition of the PCIV, a review of potential safety benefits, lessons-learned, and progress to date towards crashworthiness of PCIVs as well as proposed safety performance specifications and research needs.

Special appreciation is due to Dr. Aviva Brecher and Dr. John Brewer from the Volpe National Transportation Systems Center (Volpe Center) within the DOT Research and Innovative Technologies Administration (RITA), for providing technical support, feedback and guidance during the development of this report. This vehicle safety research project was sponsored by Stephen Summers, Chief and Sanjay Patel of the Structures and Restraints Research Division, National Highway Safety Administration (NHTSA) of the U.S. Department of Transportation (USDOT).

Additionally, the authors acknowledge the Crashworthiness Working Group of the Composite Materials Handbook, CMH-17. This group of researchers has recognized the importance of the dual-emphasis approach of standardization of crashworthiness properties and the evaluation of predictive analysis techniques. The collaboration and dissemination of knowledge within this forum has without doubt accelerated development in these research areas.

TABLE OF CONTENTS

1. **DEFINITION OF A PRELIMINARY SET OF MINIMUM PCIV SAFETY PERFORMANCE SPECIFICATIONS** 11
   1.1 INTRODUCTION 11
   1.2 THE PLASTIC AND COMPOSITE INTENSIVE VEHICLE (PCIV) 12
   1.3 POTENTIAL SAFETY BENEFITS OF PCIVS 16
       1.3.1 Introduction 16
       1.3.2 Crashworthiness and Crash Avoidance 17
       1.3.3 Potential Safety Benefits of Composite Materials 17
       1.3.4 Safety Considerations Related to the Reduced Mass of Composite Materials 18
       1.3.5 Safety Benefits Through the Use of Increased Crush Distance 20
       1.3.6 The Relative Safety Benefits of Size Versus Mass 20
   1.4 PROPOSED SAFETY SPECIFICATIONS FOR PCIVS 22
       1.4.1 Introduction 22
       1.4.2 Case for PCIV Safety Benefits 23
       1.4.3 Building Block Approach for PCIV Structural Components 23
           1.4.3.1 Level I. Coupon and Element Level 25
           1.4.3.2 Level II. Subcomponent and Component Level 27
           1.4.3.3 Level III. Sub-Assembly Level 29
           1.4.3.4 Level IV. Full-Scale Level 30

2. **LESSONS LEARNED FROM COMPOSITES IN HIGH PERFORMANCE CAR APPLICATIONS** 31
   2.1 INTRODUCTION 31
   2.2 LESSONS LEARNED 31
       2.2.1 Formula 1 31
       2.2.2 Le Mans 34
       2.2.3 High-End Supercars with Composite Safety Cells 36
       2.2.4 Research and Development Activities on Composite Vehicles 49
   2.3 CONCLUSIONS 50

3. **DEVELOPMENT OF STANDARDIZED TEST AND EVALUATION PROCEDURES** 52
   3.1 INTRODUCTION 52
       3.1.1 Candidate materials 53
   3.2 OVERVIEW OF FEDERAL MOTOR VEHICLE SAFETY STANDARDS RELEVANT TO THE DEVELOPMENT OF PCIVS 54
       3.2.1 Front and Rear Impact 55
       3.2.2 Side Impact 55
       3.2.3 Roof crush 56
   3.3 TEST AND EVALUATION PROCEDURES FOR COMPOSITE MATERIALS 56
       3.3.1 Sustained Crush Stress 56
       3.3.2 Compressive Strength 57
       3.3.3 Interlaminar Shear Strength 58
       3.3.4 Internal Damping 58
       3.3.5 Strain To Failure 59
       3.3.6 Mechanical Damage 59

3.3.7 Environmental Effects.. 60
3.3.8 Adhesives 60
3.3.9 Joint geometry 61
3.3.10 Crack Arresters at Bonded Joints 62
3.4 EVALUATION PROCEDURES FOR COMPOSITE DESIGNS AND COMPONENTS 63
3.4.1 Failure Modes 63
3.4.2 Design Evaluation 64
3.5 EVALUATION TOOLS FOR COMPOSITE DESIGNS AND COMPONENTS 64
3.5.1 Material Properties 65
3.5.2 Loading Conditions 65
3.6 TEST PROCEDURES FOR COMPOSITE DESIGNS AND COMPONENTS 65
3.7 CONCLUSIONS AND RECOMMENDATIONS 66

4. **SUMMARY OF PROGRESS IN MATERIAL DATABASES, TEST METHOD DEVELOPMENT, AND CRASH MODELING 67**
    4.1 INTRODUCTION 67
    4.2 SUMMARY OF PROGRESS: MATERIAL DATABASES 67
        4.2.1 Sources of Material Databases for Composites 68
        4.2.2 Specialized Crashworthiness Properties 69
        4.2.3 Summary: Material Databases 71
    4.3 SUMMARY OF PROGRESS: CRASHWORTHINESS TEST METHOD
        DEVELOPMENT 72
        4.3.1 Classifications of Composite Crushing 73
        4.3.2 Development of Crush Initiating Triggers 74
        4.3.3 Coupon-Level Test Methods 77
            4.3.3.1 Self-Supporting Coupon Test Methods 78
            4.3.3.2 Flat Coupon Test Methods 79
        4.3.4 Element-Level Test Methods 86
        4.3.5 Testing to Investigate Strain Rate Effects 88
        4.3.6 Summary: Crashworthiness Test Methods 89
    4.4 SUMMARY OF PROGRESS: COMPOSITE CRASHWORTHINESS MODELING 90
        4.4.1 Overview of Crashworthiness Modeling 91
        4.4.2 Categories of Crush Front Modeling 92
            4.4.2.1 Progressive Damage Modeling 93
            4.4.2.2 Continuum Damage Mechanics Modeling 100
            4.4.2.3 Multi-Scale Modeling 101
            4.4.2.4 Phenomenological Crush Modeling 102
        4.4.3 Delamination Modeling 103
        4.4.4 Damage Modeling Away From the Crush Front 104
    4.5 OTHER FAILURE MODELS 107
    4.6 VALIDATION OF CRASHWORTHINESS MODELING 108
        4.6.1 Tube Testing 108
        4.6.2 CMH-17 Crashworthiness Working Group Round Robin 108
            4.6.2.1 Progress as of Atlanta CMH-17 Meeting, Nov 2009 111
            Shell Approach 111
            Multi-Layer Shells 112
            Multi-Layer Solid 112

    4.7   Summary: Composite Crashworthiness Modeling 113

5. RESEARCH NEEDS FOR MATERIAL DATABASES, TEST METHOD DEVELOPMENT, AND CRASH MODELING 114

    5.1   Introduction 114
    5.2   Material Databases 114
        5.2.1   Current Status 114
                Recommendations 115
        5.2.2   115
            5.2.2.1   Identification of Required Properties for Crashworthiness Databases 116
            5.2.2.2   Crashworthiness Screening Testing of Candidate Composite Materials 117
            5.2.2.3   Development of Material Database for Crashworthiness Model Development 117
        Crashworthiness Test Methods 118
    5.3   118
        5.3.1   Current Status 118
            5.3.1.1   Coupon Test Methods 118
            5.3.1.2   Element-Level Test Methods 120
        5.3.2   Recommendations 120
            5.3.2.1   Further Development and Standardization of a Flat-Coupon Composite Crashworthiness Test Method 120
            5.3.2.2   Further Development and Standardization of a Tube Test Method 121
            5.3.2.3   Testing for Material Screening and Crashworthiness Model Development 121
    5.4   Crashworthiness Modeling 122
        5.4.1   Current Status 122
        5.4.2   Recommendations 125
            5.4.2.1   Further Assessment of Modeling Approaches for Crashworthiness Modeling 125
            5.4.2.2   Assessment of Modeling Capabilities to Predict Response of the Back-Up Structure During the Crush Event 126
            5.4.2.3   Accounting for the Stochastic Nature of Crush Force Inputs and the Factored Allowables in the Back-up Structure 126
            5.4.2.4   Development of a Reusable/Universal Benchmark System(s) for Crashworthiness 127
            5.4.2.5   Revival of the ACC Focal Project 3 Whole Vehicle Crash Analysis 128
    5.5   Summary 129

6. REFERENCES 130

LIST OF FIGURES

FIGURE 1-1. BUILDING BLOCK APPROACH AS ENVISIONED FOR PCIV STRUCTURAL COMPONENTS. 25
FIGURE 1-2. SUB-COMPONENT AND COMPONENT LEVEL CONE STRUCTURES. 28
FIGURE 2-1. PORSCHE CARRERA GT FOLLOWING FLIPPING OVER SEVERAL TIMES 37
FIGURE 2-2. PORSCHE CARRERA GT FOLLOWING POLE STRIKE 37
FIGURE 2-3. FRONT VIEW OF PORSCHE CARRERA GT FOLLOWING POLE STRIKE .38
FIGURE 2-4. PORSCHE CARRERA GT FOLLOWING SIDE IMPACT 38
FIGURE 2-5. PORSCHE CARRERA GT FOLLOWING SIDE IMPACT 39
FIGURE 2-6. PORSCHE CARRERA GT FOLLOWING CRASH INTO TREE 40
FIGURE 2-7. BUGATTI EB110 FOLLOWING SIDE IMPACT 40
FIGURE 2-8. FERRARI ENZO FOLLOWING SIDE IMPACT 41
FIGURE 2-9. FERRARI ENZO "LOOSELY RE-ASSEMBLED" AFTER SIDE IMPACT 42
FIGURE 2-10. FERRARI ENZO FOLLOWING FRONTAL IMPACT 42
FIGURE 2-11. FERRARI ENZO FOLLOWING FRONTAL IMPACT. NOTE FUNCTIONING CLOSURES 43
FIGURE 2-12. FERRARI ENZO FOLLOWING MULTIPLE FLIPS AND ROLLOVERS 44
FIGURE 2-13. FERRARI ENZO, VIEW FROM OPPOSITE SIDE 44
FIGURE 2-14. MCLAREN SLR FOLLOWING MULTIPLE ROLLS FROM HIGH SPEED CRASH 45
FIGURE 2-15. MCLAREN SLR FOLLOWING HIGH SPEED FRONT OFFSET IMPACT ... 46
FIGURE 2-16. FERRARI ENZO FOLLOWING 240 KM/HR (150 MPH) CRASH 47
FIGURE 2-17. PORSCHE 911 FOLLOWING HIGH SPEED CRASH 48
FIGURE 2-18. FERRARI 360 MODENA FOLLOWING HIGH SPEED CRASH 48
FIGURE 2-19. PORSCHE BOXSTER FOLLOWING EXTREME SIDE INTRUSION 49
FIGURE 3-1. CONVENTIONAL AND PEEL RESISTANT JOINT GEOMETRY. 62
FIGURE 4-1. TYPICAL LOAD VERSUS DISPLACEMENT PLOT OBTAINED FROM PROGRESSIVE CRUSHING OF A COMPOSITE TEST SPECIMEN. 70
FIGURE 4-2. CLASSIFICATIONS OF COMPOSITE CRUSHING. 74
FIGURE 4-3. CRUSH TRIGGERS USED IN COMPOSITE SPECIMENS: (A) BEVEL, (B) STEEPLE, (C) NOTCH, (D) TULIP. 75
FIGURE 4-4. SECTION VIEW OF A STANDARD PLUG TRIGGER USED WITH A SQUARE TUBE SPECIMEN. 76

FIGURE 4-5. INITIATING CRUSH USING CLOSED-END FEATURE 77

FIGURE 4-6. CROSS SECTION OF DLR SEGMENT COUPON 78

FIGURE 4-7. CROSS SECTION OF SINE WAVE WEB COUPON 78

FIGURE 4-8. CROSS SECTION OF CMH-17 CORRUGATED COUPON 79

FIGURE 4-9. CRUSH TEST FIXTURE DESIGN OF LAVOIE ET AL. [97-99] FOR FLAT COUPONS (SHOWN CONFIGURED FOR QUASI-STATIC LOADING OF THE SMALLER SPECIMEN SIZE) 80

FIGURE 4-10. CRUSH TEST FIXTURE DESIGNS OF (A) DUBEY AND VIZZINI [101] AND (B) CAUCHI SAVONA AND HOGG [102] 81

FIGURE 4-11. CRUSHING AND TEARING OF A FLAT COUPON USING THE TEST FIXTURE DESIGN OF LAVOIE ET AL. 82

FIGURE 4-12. TEST FIXTURE DEVELOPED BY ENGENUITY LIMITED 83

FIGURE 4-13. FLAT COUPON CRUSH TEST FIXTURE DESIGNS OF: (A) TAKASHIMA ET AL. [104], AND (B) FERABOLI [105]. 84

FIGURE 4-14. FLAT COUPON TEST FIXTURE OF GARNER AND ADAMS 85

FIGURE 4-15. CHARACTERISTIC ENERGY ABSORPTION VERSUS GAP HEIGHT PLOT OBTAINED FROM FLAT-COUPON CRUSH TESTING 86

FIGURE 4-16. SCHEMATIC OF A TUBE CRUSH TEST WITH INTERNAL PLUG TRIGGER 87

FIGURE 4-17 SCHEMATIC REPRESENTATION OF A CRUSH EXPERIMENT PERFORMED ON A COMPOSITE MATERIAL 92

FIGURE 4-18. CONVENTIONAL EXPLICIT FINITE ELEMENT SIMULATION OF CRUSH EXPERIMENT 95

FIGURE 4-19. EFFECTS OF FILTERING RESULTS FROM COMPOSITE CRASH SIMULATIONS 97

FIGURE 4-20. CONVENTIONAL EXPLICIT FINITE ELEMENT SIMULATION OF CRUSH EXPERIMENT 99

FIGURE 4-21. TRACTION-SEPARATION RESPONSE OF INTERFACE ELEMENTS USED FOR DELAMINATION MODELING 104

FIGURE 4-22. RESULTS OF TESTING AND ANALYSIS OF CARBON COMPOSITE COMPLEX CONE SUBJECTED TO IMPACT 106

FIGURE 4-23. PREDICTED FAILURE PROGRESSIONS FROM COUPON TESTING 110

INDEX OF TABLES

TABLE 4-1. SUMMARY OF MATERIAL DATABASES FOR COMPOSITE MATERIALS 72

TABLE 4-2. SUMMARY OF CRASHWORTHINESS TEST METHODS 90

TABLE 4-3. CRASHWORTHINESS LOAD CASES CONSIDERED FOR AUTOMOTIVE DEVELOPMENT 91

EXECUTIVE SUMMARY

The Plastic and Composite Intensive Vehicle (PCIV) has the potential to revolutionize the automotive sector, due to the inherent benefits of these materials. Composite materials provide high strength-to-weight and stiffness-to-weight ratios as well as excellent energy absorbing capability per mass. However, the use of these materials in automotive structures requires an in-depth knowledge of their unique performance characteristics in the crash and safety environment.

This report attempts to identify outstanding safety issues and research needs for future PCIVs in order to facilitate deployment of safe PCIV vehicles by 2020. Specific objectives of this report are to:
- Propose a definition of a PCIV
- Define a preliminary set of minimum PCIV safety performance specifications
- Develop approaches and metrics for the characterization and quantification of potential safety benefits of automotive plastics and composites
- Develop objective test and evaluation procedures for materials, designs and components of emerging PCIV concepts, to ensure compliance with Federal Motor Vehicle Safety Standards (FMVSS)
- Summarize progress and provide recommendations for future research in materials databases, test method development, and crash modeling.

Included in this report are the following items of significance towards addressing these objectives:
- A dual-component PCIV definition is proposed, which includes requirements on the "areal density" as well as the weight percentage of plastics and composite materials to ensure that the weight and efficiency objectives are prerequisite.
- Potential safety benefits of automotive plastics and composites are reviewed and safety performance specifications for PCIVs are proposed.
- The Building Block approach envisioned for PCIV structural components is reviewed. Proposed safety specifications associated each level of the Building Block are identified. Future research efforts required to develop such safety specifications are identified.
- Lessons learned from the racing industry and from limited production, high-performance supercars with extensive use of composite materials are summarized.
- Changes and additions to test and evaluation procedures due to PCIVs are discussed, with a focus on ensuring their compliance with FMVSS.
- Progress is summarized in three topic areas pertinent to crashworthiness of PCIVs: material databases, crashworthiness test method development, and crash modeling.
- A summary of the current status and research needs is presented in material databases, crashworthiness test method development, and crash modeling.

1. DEFINITION OF A PRELIMINARY SET OF MINIMUM PCIV SAFETY PERFORMANCE SPECIFICATIONS

1.1 INTRODUCTION

Plastic and Composite Intensive Vehicles (PCIVs) have the potential to revolutionize the automotive sector, due to the inherent benefits of composite materials. However, the behavior of composites in the crash and safety environment requires an in-depth knowledge of the materials and their unique performance characteristics. In metallic structures, plastic deformation is the primary failure mode associated with energy absorption during a crash event. In composites, however, energy absorption is often associated with brittle-type fractures, resulting in the destruction and disintegration of the structure in the crush zone. Regardless of the type of material used, it is the formation and propagation of these high energy absorbing failures in a crush zone while maintaining the structural integrity of the remaining structure away from the crush front that leads to a crashworthy structure. Both of these attributes need to be present.

The goals of this research report are to address outstanding safety issues for future Plastic and Composite Intensive Vehicles (PCIVs), in order to facilitate deployment of safe PCIV vehicles by 2020. Specific research objectives are to:
 a) Propose a definition of a PCIV
 b) Define a preliminary set of minimum PCIV safety performance specifications
 c) Develop approaches and metrics for the characterization and quantification of potential safety benefits of automotive plastics and composites
 d) Develop objective test and evaluation procedures for materials, designs and components of emerging PCIV concepts, to ensure compliance with FMVSS
 e) Summarize progress and provide recommendations for future research in materials databases, test method development, and crash modeling.

In this chapter, the concepts of the Plastic and Composite Intensive Vehicles (PCIVs) are discussed and an expanded definition of the PCIV created. Based on this definition, safety performance specifications for PCIVs are proposed. Finally, the potential safety benefits of automotive plastics and composites are reviewed, with an emphasis on the current status and future directions in characterizing and quantifying such safety benefits.

1.2 THE PLASTIC AND COMPOSITE INTENSIVE VEHICLE (PCIV)

In fiscal year 2006, the United States Congress directed the National Highway Traffic Safety Administration (NHTSA) to undertake research based on the broad application of plastics and composites in the automotive industry based on the following recommendation:

> ***Plastic and Composite Vehicles*** *-- The Committee recognizes the development of plastics and polymer-based composites in the automotive industry and the important role these technologies play in improving and enabling automobile performance. The Committee recommends ($500,000) to continue development of a program to examine possible safety benefits of Lightweight Plastic and Composite Intensive Vehicles [PCIV]. The program will help facilitate a foundation between DOT, the Department of Energy and industry stakeholders for the development of safety-centered approaches for future light-weight automotive design"* [1].

To date, there has been no accepted definition of a Plastic and Composite Intensive Vehicle (PCIV). The word "intensive" in PCIV suggests that plastics and composites should compose a significant portion of the vehicle. However, some major vehicle components, such as the engine block and power train, are not viewed as candidates for plastics and composites. As a result, the definition of plastic and composites "intensive" is intended primarily for other vehicle components.

The creation of a definition of a PCIV was a subject of discussion at the 2008 Safety Characterization of Future PCIVs Workshop [2]. The discussion focused on establishing a definition based on either a volume or weight percentage of plastics and composites within a vehicle. Among the ideas discussed for defining a PCIV was excluding the engine block and power train as well as requiring that 30% to 40% of the weight of one or more automotive subsystem be composed of plastics and composites. Currently, a typical passenger vehicle's weight consists of approximately 10% plastics by weight, whereas a majority of the vehicle's weight (greater than 75%) consists of steel [2].

NHTSA in their report concentrating on the safety-related research issues affecting the deployment of PCIVs in 2020 [3, 4] attempted to refine the definition further. Referencing the earlier industry experts workshop recommendations, the authors indicated that the qualification criteria from the OEMs and material suppliers would be a minimum of 30% to 40% (by weight) plastics and composite content in one or more subsystems beyond interior trim. The authors highlighted that this criteria was less stringent than the DOE/USCAR light-weighting "Factor of Two" goal desired for improved fuel efficiency.

Subsequently the Plastics Division of the American Chemistry Council (ACC) adopted the lower bound of this threshold in their definition, or 30% (by weight) of lightweight plastics and composite content in one or more subsystems beyond interior trim. Once again, the definition of the subsystems was not expanded upon. Also, the requirement for making the vehicle lighter weight than current steel cars was not stated, although this was clearly part of their overall vision for the PCIV [5].

Although it is clear that all the contributors to the definitions have assumed the vehicle mass will be reduced for a PCIV, a situation could exist whereby a vehicle qualifies for PCIV status through the addition of plastic and composite mass to a vehicle subsystem up to the 30% threshold with no attempt at weight reduction.

While the research directive from Congress was clear in its research directive by prefacing the subject "Lightweight PCIV" the word *lightweight* has been omitted from the published definitions to date. However, the primary motivation behind the PCIV initiative, significant vehicle weight reductions with maintained or improved vehicle safety, must be taken into account when defining a PCIV. Cheah et al. [6] have postulated that vehicle weight reduction can be brought about through simple material substitutions, redesign of existing vehicles, and reducing vehicle size. However, the authors note that weight reductions can be accomplished using materials other than plastics and composites (ex: replacing steel components with aluminum). Thus the use of a weight reduction metric to define a PCIV by itself does not provide the assurance that the vehicle will be plastics and composites *intensive.*

Similarly, fuel efficiency may be considered as a metric in the definition of a PCIV. Current indications suggest that future vehicle development will be based increasingly on increases in fuel mileage and reducing the negative effects of vehicles on the environment. Fuel efficiency increases can be realized through vehicle weight reductions, which may involve intensive usage of plastics and composites. However, additional fuel efficiency may be achieved from other changes that do not affect the use of plastics and composites, including decreased engine size and increased engine efficiency. Thus similar to weight reduction, the use of a fuel efficiency metric by itself does not provide the assurance that the vehicle will be plastics and composites intensive.

While reductions in vehicle weight and increases in fuel economy are viewed as primary motivations behind the PCIV initiative, these future vehicles must also be competitive in terms of safety. As such, it can be argued that the definition of a PCIV needs to include safety considerations. For example the PCIV definition could require that the occupants of a PCIV need to be as safe in a collision with an existing conventional vehicle of the same class as they would be if the collision occurred between two conventional vehicles. While this "safety equivalence" requirement may not provide a near-term safety benefit, it is expected to provide a longer term safety advantage as the fleet migrates to lighter vehicles: a PCIV will be safer in a collision with another comparable PCIV of the same class.

An additional consideration in defining a PCIV involves the use of fundamental design concepts for plastics and composites. Rather than simple material substitutions of plastics and composites into existing metallic designs, components or entire assemblies of a vehicle could be redesigned specifically to exploit the advantages of plastics and composites. The number of significant vehicle components that are designed for plastics and composites could be a consideration in the definition of PCIV. However, such a definition would be difficult to quantify.

The first component of the proposed PCIV definition is as follows:

The areal density of a PCIV must be less than 120 kg/m^2 (0.17 lb/in^2).

This areal density was obtained by the authors based on the North American PNGV (Partnership for a New Generation of Vehicles) data (mass = 900 kg, length = 4.75 m, and width = 1.8 m), resulting in an areal density of 104 kg/m^2 (0.148 lb/in^2). In contrast, current vehicles have an areal density of approximately 168 kg/m^2 (0.239 lb/in^2)

The second component of the proposed PCIV definition focuses on the percentage of the vehicle weight that consists of plastics and composites. When proposing such a definition, at least six categories of vehicle subsystems may be considered:

1. <u>Body components (including closures)</u>. Currently, body structures are roughly 95% metallic, and are likely candidates for plastic or composite replacement.
2. <u>Chassis components (steering, suspension, and wheels)</u>. Currently, vehicle chassis are roughly 95% metallic, and are considered reasonable candidates for plastic or composite replacement.
3. <u>Interior trim (including cross-car beam)</u>. Currently, interior trim is composed of approximately 80% plastics except for steel seat structures and the cross car beam, which can account for roughly 50% of the total mass of the interior trim.
4. <u>Exterior trim</u>. Currently, approximately 80% of the exterior trim is non-metallic.
5. <u>Engine</u>. Currently, the engine is greater than 95% metal and is an unlikely candidate for significant weight savings through the use of plastics and composites.
6. <u>Transmission</u>. Similar to the engine, the transmission is greater than 95% metal and is an unlikely candidate for significant weight savings through the use of plastics and composites.

Of the six vehicle subsystems described above, the engine and transmission are not considered candidates for plastics and composites and thus should not be included in defining a PCIV. In contrast, the body and chassis are believed to be the greatest source of "convertible mass" – from metallics to plastics and composites. Thus the second component of the proposed definition of a PCIV, based on the weight percentage of plastics and composites, is as follows:

A PCIV must meet one or more of the following requirements:
- *Greater than 80% plastics and composites by weight in either the body or chassis*
- *Greater than 50% plastics and composites in the combined weight of the body and chassis*
- *Greater than 55% plastics and composites in the combined weight of the body, chassis, and interior trim*

A possible complication when considering the definition of a PCIV is the different vehicle classes, ranging from subcompact cars to Sport Utility Vehicles (SUV's) and vans. Of concern is that a single definition, such as the one above, may be well suited for some vehicle classes while leading to contradictions in others. A potential contradiction or violation of the PCIV objective would be a situation where a vehicle qualifies as a PCIV through the addition of non-structural composites to a conventional vehicle design, leading to a heavier and less efficient design than the base vehicle.

The above definition was developed by the authors based on the ACC Focal Project 3 baseline, which in turn was based on the North American PNGV (Partnership for a New Generation of Vehicles) class cars that are described as:
- 5/6 passenger sedan
- 3.115 cubic meter (110 cubic feet) interior volume (including passenger and luggage space)
- Curb weight of 907 kg (2000 lb)
- Capable of achieving 34 km/liter (80 miles per gallon) fuel economy.

The reference vehicle was the Chrysler "Cloud" Cars or JA Series; Dodge Stratus, Chrysler Cirrus and Plymouth Breeze. In the above definition, the percentage content of plastics and composites needs to be carefully evaluated not to generate misplaced classifications. It is recommended that a series of case studies are constructed and peer reviewed to ensure the majority of stakeholders in the field are in agreement of the PCIV classifications for each vehicle class.

There are two factors to consider when embarking on a project to dramatically save weight in a vehicle. The first involves the direct savings associated with the advanced design and development using the plastic and composite materials. The second involves the savings resulting from the overall vehicle mass being reduced.

Weight savings due solely to reducing the vehicle mass was successfully demonstrated during the development of the 1992 Honda FireBlade motorcycle, without the need for deployment of plastics and composites in the primary structures. Upon completion of the first model of the FireBlade, the originator and designer allegedly sent the engineering team back to redesign every component in light of the weight savings achieved by other motorcycle designers. The new lightweight pistons resulted in lighter connecting rods, etc. The resulting design, without the use of alternative materials, was a 20 percent saving over the lightest competitor and a market-leading position on weight which would not be matched for another six years by the competition [7].

The direct mass reduction resulting from the deployment of plastics and composites in the Body In White (BIW) structure may be of the order of 60 percent of the equivalent steel structure. All other structures and systems left unchanged, this would only result in a 15 percent overall mass reduction. Discounting the opportunity of deploying plastics and composites elsewhere in the vehicle, however, the engine and powertrain now have to provide and endure significantly less in order maintain the performance attributes of the vehicle. As a result, both the engine and powertrain can be correspondingly downsized. This presents further opportunities for weight reduction as the vehicle mass spirals down and is entered back into the design process to further yield savings across all systems [8].

The weights saving opportunity for a vehicle designed from the outset in composites and with a view towards fuel economy are immense. Applying plastics and composites to key structures of the vehicle where direct savings are possible reduces the need to apply the material to components which are not ideally suited, and as a result presents a higher chance of successful conversion.

1.3 POTENTIAL SAFETY BENEFITS OF PCIVS

1.3.1 Introduction

The principal benefit for using plastics and composites in automotive structures is believed to be the opportunity for weight savings in the mass production of future automobiles. However, a number of safety benefits have been identified for composites, including the high Specific Energy Absorption (SEA) and specific strength being translated into the ability to prevent intrusion. The PCIV is a relatively new concept, and has not been developed extensively to the point of prototype evaluation. Thus, the potential safety benefits of PCIVs have yet to be fully demonstrated. This section addresses the safety considerations related to the development of PCIVs – both the potential safety benefits as well as the negative attributes of many current plastic and composite architectures that must either be eliminated or overcome.

As discussed at the 2008 PCIV Workshop held at the Volpe Center [2], the safety benefits of plastics and composites can be divided into two general classifications based on their usage. The first classification includes structural components that may be used to absorb energy during an impact, either with another vehicle or with a stationary object. For such structures, the property that is being exploited using composites is their high SEA, or energy absorption per unit mass. A second classification involves parts and components that are non-structural in nature, and are used primarily in the interior of the automobile to reduce impact forces imparted on vehicle occupants during a crash. These parts and components are currently placed on conventional steel body/chassis automobiles, and cannot be considered as a distinguishing feature of PCIVs. This section will focus on the safety benefits associated with the use of plastics and composites in the vehicle structure.

To better understand the safety threats to a vehicle occupant during a crash, it is useful to consider the sequence of "collisions" that occur. In the first collision, the vehicle strikes another vehicle, hits an object, rolls over, or experiences a combination of any of these events. In this initial collision, the vehicle's exterior is partially crushed. As kinetic energy of the impacting mass or masses is absorbed through crushing, the remaining kinetic energy may be shared among the interacting bodies, which has the potential to reduce speeds. Depending on the crash severity, the location and direction of crash force, and the stiffness of the vehicle and of the impacted object, crushing can result in components intruding into the vehicle's passenger compartment, a potential source of occupant injury. In the second "collision", occupants strike interior surfaces of the vehicle and passenger restraints as vehicle deceleration occurs. Restraints such as seat belts, air bags, and perhaps padding are used to reduce injuries from this type of collision. In general, such collisions between a vehicle occupant and either restraints or intrusions tend to occur when the crash involves a high degree of kinetic energy or an unfavorable crash geometry. The third "collision" involves the impacts that occur among parts of the occupant's body, such as organs and skeleton. If the contact between an occupant and a hard surface is brief enough, the interaction among parts of the body may be small and the occupant may avoid injury.

1.3.2 Crashworthiness and Crash Avoidance

It is useful to distinguish between the two general characteristics of vehicles that protect their occupants from death or serious injury in a crash: crash avoidance and crashworthiness. Crash avoidance is the ability of a vehicle, through driver-controlled as well as automatic handling and braking, to avoid a serious crash altogether, braking distances of vehicles have been regulated and tested for a number of years, other means of crash avoidance such as Electronic Stability Control (ESC) are now subject to regulation as their effect on reducing fatalities is significant [9]. Consumers Union, the non-profit publisher of Consumer Reports [10] conducts handling and braking tests on vehicles. Crashworthiness refers to the ability of a vehicle to protect its occupants once a crash has occurred. Under the New Car Assessment Program (NCAP), the National Highway Traffic Safety Administration (NHTSA) conducts crash tests in a laboratory setting to ensure that new vehicles comply with crashworthiness standards. Based on the results of NCAP testing, a rating of up to "5 stars" is assigned to each vehicle model, and made publicly available on the NHTSA website [11]. Currently, NHTSA conducts tests of frontal and side impact, as well as rollover crashes for the NCAP program. Additionally, the Insurance Institute of Highway Safety (IIHS) conducts these and other tests, and publishes their results on their website [12]. When these tests were first introduced in 1979, no vehicles received a 5-star rating on the frontal impact test, and many vehicles received only 1-star or 2-star ratings. In contrast, nearly all of the newest model year cars now earn 4-star or 5-star ratings from the frontal impact test [13].

1.3.3 Potential Safety Benefits of Composite Materials

Composite materials possess many material properties and characteristics that differ significantly from those of conventional metallic materials such as steel and aluminum. When considering the potential safety benefits arising from using composite materials in structural components of a vehicle, two material-related safety benefits may be identified: improvements in SEA and added resistance to intrusion.

The most commonly presented safety benefit of using composites in vehicle structural components is the possibility of higher SEA than available with metallic materials such as steel and aluminum. In metallic structures, energy is absorbed through plastic deformation as the structure is folded in an accordion manner. In contrast, the mechanism by which composite materials absorb energy most efficiently is through material fragmentation, such that the composite material disintegrates along a crush front as crushing progresses. The level of fragmentation, corresponding to the fineness of the debris created, determines the level of energy absorption. One widely quoted source of comparison data lists the SEA for carbon thermoset composites at more than 100 kJ/kg (33.5×10^3 ft-lb/lb), compared to an SEA of approximately 30kJ/kg (10.1×10^3 ft-lb/lb) for aluminum and 20 kJ/kg (6.7×10^3 ft-lb/lb) for steel [14]. Through their own experience, the authors can verify SEAs in excess of 80 kJ/kg (26.8×10^3 ft-lb/lb) for some of the better performing carbon thermoset systems and 40 kJ/kg (13.4×10^3 ft-lb/lb) for some glass thermoset systems.

The widely quoted graphic from Hermann et al. [15], which illustrates the relative SEA of composites and competing metals, forms a basis of justification for the use of composite materials as efficient energy absorbers. While the authors defend the premise that composites are highly efficient for energy absorption, the SEA values quoted in this widely reproduced graphic are not believed to have been obtained using a consistent test velocity, specimen geometry or test method, and real-world automotive applications would fail to deliver the inferred performance. For example, the SEA value of 250 kJ/kg (83.9 x 10^3 ft-lb/lb) for carbon/thermoset materials is believed to be related to a quasi-static test of a short 55 mm (2.2 in.), 55 mm (2.2 in.) diameter tube with a 2.67 mm (0.105 in.) wall thickness at an applied displacement rate of 16.7 microns per second (6.6 x10^{-4} in./sec) [16]. Additionally, the SEA value provided for honeycomb is not consistent with values obtained for commonly used aluminum honeycombs. A common 3.2 mm (1/8 in.) cell aluminum honeycomb at 130 kg/m^3 (8.1 lb/in^3) yields a crush stress of 40kJ/kg (13.4 x 10^3 ft-lb/lb) [17]. These points highlight the need for standardized methods of assessing the performance of candidate materials.

Another possible safety benefit of using composite materials in vehicle structural applications is their resistance to intrusion during a crash event. While object intrusion into the passenger compartment is a concern in any vehicle crash, it is of particular concern for single-vehicle crashes involving objects such as trees and poles as well as vehicle roll-overs. According to 2008 crash fatality data from the NHTSA Fatality Analysis Reporting System (FARS), 46% of fatal crashes, and 52% of occupant fatalities, are single-vehicle events, with the vehicle either crashing into an object or rolling over [18]. Further, 22% of all traffic fatalities (and 25% of all vehicle occupant fatalities) in 2008 came from vehicle rollovers, either as a first or subsequent event. The prevention of object intrusion during a crash event is an important safety consideration. Composite materials have high specific strengths (strength-to-mass ratios) which allows for composite structural sections such as the A-pillar, header and cant rails to be made larger with disproportionately increased section strength and stiffness, a safety advantage. As a result, maximum safety cell load levels (prior to collapse) typically are higher threshold than conventional metallic vehicle structures in the domain where the resulting accelerations on the occupants are survivable.

1.3.4 Safety Considerations Related to the Reduced Mass of Composite Materials

Currently there is some debate regarding the role of vehicle mass in crashworthiness and vehicle safety. Lighter weight vehicles are often thought to be less safe when involved in a collision with a heavier vehicle. However, a more detailed thought experiment of such a collision reveals that mass is not the only consideration in vehicle safety. Each vehicle enters the collision with kinetic energy, $KE = \frac{1}{2} mv^2$, where m is the vehicle mass and v is the velocity. Assuming that the velocity of both vehicles goes to zero as a result of the collision, then the sum of the kinetic energies of the impacting vehicles must be absorbed by the two vehicles. The question that arises is how much energy will be absorbed by each vehicle? The answer has more to do with the force required to produce crushing than the mass. That is, a large vehicle with a strong and heavy crush structure will not begin to crush until a relatively high crush force is produced. If the lighter vehicle is equipped with a lighter crush structure that begins crushing at a lower crush force, then the crush structure in the lighter vehicle will experience crushing first. As the lighter vehicle's crush structure reaches the end of its crush length and the forces subsequently increase, further crushing will occur in the crush structure of the larger vehicle, and further energy will be absorbed. Thus while the mass of the vehicles influences the crash energy that must be absorbed, the design of the crush structure of each vehicle, including the force required to produce crush and the total energy absorption capacity of the structure, determine the progression of crush during the collision and the resulting decelerations of the two vehicles.

For head-on collisions when there is a substantial difference in mass between the two impacting vehicles, increased mass generally offers additional protection of vehicle occupants. As described above, however, differences in the design of the crush structures between the two vehicles are perhaps of greater importance than the actual difference in mass. The crashworthiness performance of an automobile today of a given mass is considerably improved from an equivalent mass automobile from a few decades below. This has been exemplarily demonstrated by the Insurance Institute for Highway Safety in their recent frontal offset crash test crash of a 1959 Chevrolet Bel Air with a 2009 Chevrolet Malibu that is 74 kg (163 lb.) lighter [19]. Following the collision, the occupant compartment of the 2009 Malibu remained intact whereas the one in the 1958 Bel Air collapsed. Additionally, the difference in frontal heights of impacting vehicles, particularly between cars and large Sport Utility Vehicles (SUVs) and light trucks is an important factor in head-on collisions.

Side impacts result in additional safety threats to vehicle occupants because of the minimal available crush space in the side structure of a vehicle. While the mass of a side-impacting vehicle directly affects the kinetic energy that must be absorbed, the compatibility of the impacting vehicle, both in terms of vehicle height as well as the stiffness and crush force of the impacting vehicle are also of high importance. More detailed testing is required to investigate which characteristics are most important in defining safety risks from side impacts. Greater reinforcement of the occupant compartment and installation of side curtain airbag offer increased protection to such side impacts. The key parameters affecting the prevention of intrusion under such impacts are the strength of the passenger compartment and the height and crush strength of the colliding object.

Consider, however, the case of a vehicle impacting a stationary object. The heavier the vehicle is, the greater the kinetic energy that must be absorbed by the vehicle's crush structure. Thus, higher mass vehicles require crush structures with greater energy absorption capacity to produce the same level of safety as lower mass vehicles. For any vehicle mass, an increase in the crush distance provides additional protection to the vehicle occupants, as discussed in the following section.

Off-angle impacts require further consideration during the design of the vehicle front end. In scenarios where off-angle impacts are anticipated, conical structures provide progressively increasing cross section and therefore increasing crush resistance. Further, the significance of the off-angle loading is reduced through the crushing of an increasingly large cross section. Asymmetric impacts and front pole strikes can also present similar issues which require careful consideration to ensure that the forces generated in the cross car components do not prevent the energy absorbers on the other side of the vehicle from functioning.

1.3.5 Safety Benefits Through the Use of Increased Crush Distance

As described above, components of a vehicle must experience crushing to absorb energy in a vehicle crash. To absorb energy during a frontal impact, vehicles incorporate crush structures in the space between the front of the vehicle and the passenger compartment. The amount of energy that can be absorbed by the crush structure can be thought of as the energy absorption *per unit crush length* times the available crush distance. This expression suggests that to achieve a desired level of energy absorption, the required crush distance must be chosen to account for the material used (SEA) and the design of the crush structure. In addition to the available energy absorption, occupant safety in a vehicle crash is dependent on the crush distance that is utilized. In a crash, the level of deceleration that the vehicle and its occupants will experience is dependent on the utilized crush length. Shorter crush lengths will produce greater levels of deceleration, and consequently greater safety risks to the vehicle occupants. In fact, all vehicles, regardless of size and weight, require a similar crush distance in order to decelerate the vehicle occupants at a safe level. Thus, the available crush distance in a vehicle is an important consideration in a frontal impact. It is important to note that using materials with higher energy absorption capacity (such as composite materials) to absorb more energy in a reduced crush distance will produce higher level of decelerations, and therefore greater safety risks to vehicle occupants. The usage of materials with higher SEA should not be considered as a means of reducing the crush length.

When using metallic crush structures, roughly one-third of the original length of the original structure is not crushable, as the structure is folded in an accordion manner to form a pleated column. In contrast, the mechanism by which composite materials absorb energy most efficiently is material fragmentation, such that the crush structure disintegrates as crushing progresses. The level of fragmentation, corresponding to the fineness of the debris created, determines the level of energy absorption. As a result of this failure mechanism, the usual crush distance may be a greater portion of the overall length of the crush structure when using composite materials.

1.3.6 The Relative Safety Benefits of Size Versus Mass

The link between vehicle mass and safety has been much debated over many years, and it is difficult to deconstruct the fatality data into an unequivocal position on the subject. Vehicles have consistently become heavier and larger over the decades, and a summary of this "weight spiral" for compact cars has been compiled by the European Aluminum Association [20] in response to high crashworthiness requirements and expectation by the consumer for refinement and luxury in the cabin. Automotive engineers have striven to meet the regulatory demands for the vehicle, and undoubtedly the efficiency and abilities of the current conventional vehicles are considerably in excess of their lighter weight predecessors. However, it is not reasonable to make mass the single defining factor for safety. For example, the proposition of simply increasing the mass of an early 1980's vehicle and expecting it to perform as well in the current impact test scenarios as an equivalent size and weight vehicle of today is irrational.

If restricted to using conventional metallic materials in the front crash structures of a vehicle, it is difficult to counter the argument that increasing size (crush distance and cabin space), with an increased body mass will yield a safer vehicle. As a result, the spiral of increasing vehicle weight continues. However, it is the opportunity to use materials with a higher SEA and increased effective crush distance (due to less stack-up) that can break this relationship and truly yield increased safety with reduced weight.

The American Chemistry Council published an interesting review on the general research conducted in this area and Krebs [21] also highlights the 2004 introduction of the Jaguar XJ8, which was approximately 180 kg (400 lb) lighter than the model it replaced. Although lighter, this vehicle was larger in key dimensions and able to attain the top safety rating in its luxury class. Such achievements were made through the use of aluminum, which provides only small improvements in SEA over steel. The use of composites in the vehicle crash structures with their associated significant increase in SEA, offer the opportunity of improved passive safety and also better dynamic response to improve crash avoidance.

Finally, consider the role of vehicle mass on both the avoidance of and crashworthiness during a rollover crash. Neither crash avoidance nor crashworthiness are improved due to their increased mass. In fact, heavier vehicles, such as SUVs and trucks, generally are more likely to roll over than lighter weight passenger cars due to the increased height of their center of gravity.

The fact that smaller vehicles are more crash involved has been attributed to factors such as the lack of visual presence to other drivers and increased risk taking due to their increased maneuverability. However it does not follow that future PCIVs will have increased maneuverability by virtue of reduced mass, as the further weight savings from aspects such as smaller tire size, harder tread and stiffer construction for improved rolling resistance will also balance the inherent improvements in maneuverability. Overall, this will allow automakers to design vehicles that may have similar characteristics to the mainstream vehicles of today, which although not more dynamically capable, will not increase the perception of capability to the detrimental effect of increased risk taking by the driver.

The PCIV is a future class of vehicle that will be lightweight, of standard size, and built for economy. Forecasting the interaction with future drivers is difficult. The advances in vehicle dynamic capability over recent years are not generally exploited by drivers on a regular basis. Drivers only require maneuverability when it is necessary to avoid an accident.

The safety benefits resulting from the use of plastics and composites in body structures can perhaps be best demonstrated when considering alternative material options to reduce mass. Through the use of steel, automakers have optimized structures for both energy absorption and intrusion resistance for the passenger safety cell for cases of front, offset, side and roof crush. Material thicknesses are regularly below 1 mm (0.04 in.) for various pressings in the A-pillar, cant and header rails. In order to produce a 50% weight saving in body mass for these optimised steel structures, the only realistic opportunity is to reduce part thickness further. Keeping the section at similar size would result in thicknesses reduced to 0.25 mm (0.01 in.). If the section size were increased for increased section stiffness, the thicknesses would have to be reduced even further. Out-of-plane stability will be diminished significantly, and local damage from roadway debris would reduce the stability even further. Alternatively, composite materials under consideration for automotive structures have a density of approximately one-fifth of steel. As a result, further weight reduction can come with an increase in thickness and/or an increase in section properties. In many cases, the ability to tailor the material properties in response to the required loading directions as well as the resulting improvements in damage tolerance and section stability can be significant.

1.4 Proposed Safety Specifications for PCIVs

1.4.1 Introduction

From the perspective of governing safety regulations, future PCIVs must be as safe as conventional vehicles. However, a primary motivation for PCIVs is vehicle weight reduction, leading to increased fuel efficiency. Hence an additional challenge for future PCIVs is to meet or exceed future safety regulations while reducing vehicle weight.

An important difference exists between performance *standards* and component-related *specifications* as related to vehicle safety. As discussed by Marino [22], vehicle safety regulations may be based on either performance *standards* or *specifications* on particular components. However, performance *standards* generally focus on the outputs of a prescribed test whereas *specifications* focus on the requirements of specific materials or components.

Current vehicle safety performance regulations are included in the U.S. Federal Motor Vehicle Safety Standards (FMVSS). Within the "200-series" standards related to crashworthiness, numerous standards exist that focus on safety requirements related to interior components (ex: seats, seatbelts, child restraints, head restraints, steering wheel, dash board). Additionally, several of the crashworthiness standards address vehicle safety requirements when subjected to various crash scenarios. In particular, three of these standards will be applicable to structural components of future PCIVs and may be viewed as requirements for plastic and composite structural components. These three standards are FMVSS 208 [23] for frontal impact, FMVSS 214 [24] for side impact, and FMVSS 216 [25] for roof crush resistance. Future PCIVs, which utilize plastic and composite intensive structural components, will be required to comply with these safety standards.

In contrast, to such safety performance regulations, vehicle safety specifications focus on the safety requirements of specific materials or components. The development of such specifications for plastic and composite structural components will likely be different than for conventional metallic structures, even though both must produce compliance with the appropriate vehicle safety standards. Thus PCIV-specific safety specifications for particular structural components are beneficial towards the development of PCIVs.

At present, no PCIV-specific safety specifications are known to exist. However, the development of safety performance specifications for PCIVs is viewed as helping to guide the development of composite structural components. The advanced composite materials community has formally developed a "Building Block" approach for the design of composite structures, and this approach can be viewed as suitable towards the development of plastic and composite intensive automotive vehicle components. Such a Building Block approach has been adopted by the Crashworthiness Group of CMH-17 [26] in their initial attempts to address crashworthy composite structures. This process is used to integrate both testing and analysis of structures though levels of increasing complexity. The steps or levels involved in the building block approach progress from coupons and relatively simple structural elements to subcomponents/components, and finally the entire vehicle. The following section summarizes the general usage of the building block approach towards the development of PCIV structural components, and opportunities for the development of PCIV-specific safety performance specifications at each level of development.

1.4.2 Case for PCIV Safety Benefits

In addition to regulatory standards and specifications, there is a perceived need to prove the general case for improved safety through the use of PCIVs. Government, industry, and consumers are can all be skeptical about the intensive adoption of "plastics" in vehicles for structural applications. As a result, it is important to demonstrate the enhanced safety opportunities that a PCIV will bring. A favored approach for such demonstration is the Building Block approach, to be described in the following section. This approach, if adopted in the prototype development process, may be used to demonstrate the subsequent safety benefits of the PCIV.

Whether or not the Building Block approach needs to become a formal specification or simply be maintained as good working practice within the industry is debatable. It should be noted, however, that the Building Block approach is widely used today in the development of conventional steel vehicles. As maturity of certain design, analysis, and manufacturing methods have increased, the number and size of the increments has been reduced. This is expected to be the evolutionary process that will be followed in the development of composite structures.

1.4.3 Building Block Approach for PCIV Structural Components

Although used in the aircraft and automotive industries well before the usage of composite materials (and still used in practice today), the Building Block approach has been widely accepted by the aerospace composites industry. This approach is generally viewed as of great importance for composite structures due to the lack of knowledge into the possible failure mode or modes that must be understood and considered in the design process. The steps involved in a general Building Block approach progress to address increasing structural complexity. The approach suggested by the Crashworthiness Group of CMH-17 follows the general approach followed the aerospace composites community and discussed in detail in the Composite Materials Handbook, CMH-17 [27], and has been successful deployed in the development of CZone [28]. The approach involves a mixture of testing and analysis, both of which are viewed as necessary. For most components, testing-only approaches are prohibitively expensive and do not lead to a thorough understanding of the mechanisms at play in the success or failure in a given test. Current computational analysis techniques for predicting crush performance and crashworthiness are either in their early days and as a result limited in their validated commercial deployment or in many cases still in the early developmental stages and require some level of experimental calibration and/or validation. A combined approach that utilizes testing and analysis on structures of increasing complexity is viewed as both the most efficient and the most successful for the design of crashworthy composite structures.

Figure 1-1 shows a schematic of the Building Block approach as envisioned for PCIV structural components. The building block is often drawn as a pyramid to indicate that the amount of testing to be performed decreases with increasing level of complexity as one progresses from the base level "up" the pyramid. The results from the previous level(s) of the Building Block approach are used to assist in defining aspects of the experiments and validate the computational models in the current step. Once the modeling approach is able to provide predictions with an acceptable degree of accuracy, the process can move to a higher degree of complexity associated with the next level. Variations are common when applying the Building Block approach to different components or applications, especially at the component level, at which point the number and type of test performed can vary significantly.

Another important aspect of the Building Block approach is the feedback of test results "down" the pyramid, such that comparisons may be made with data from lower levels of testing from which the design of the higher level test articles were based. Through such comparison, the performance of the higher-level test articles may be assessed and will become part of the designers' knowledge base. It is through such comparisons and assessments that confidence may be developed in moving up the levels of the Building Block.

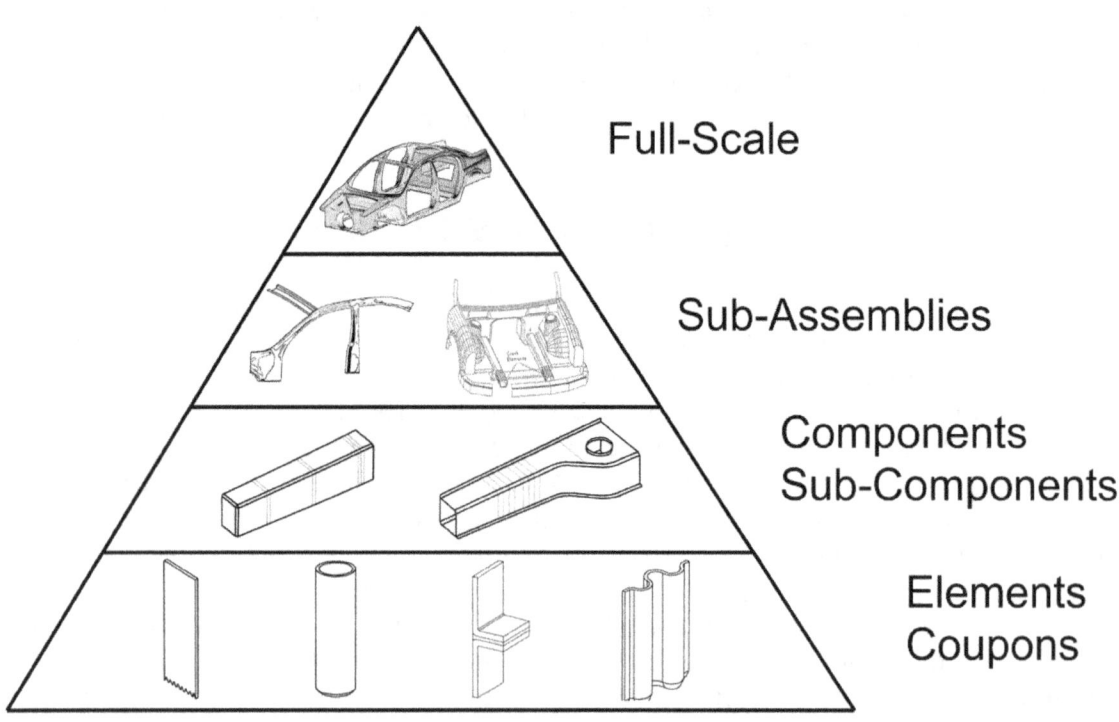

Figure 1-1. Building Block approach as envisioned for PCIV structural components.

As the Building Block approach is followed for the design of structural composite components for PCIVs, the process may be guided by component-specific safety specifications that are in agreement with the identified levels. Such safety specifications can be developed to focus on testing involved, or the predictive capabilities of modeling methods under consideration. In the following sections, each level of the Building Block approach is applied towards the development of composite structural components for future PCIVs and proposed safety specifications associated with the particular level are identified. Additionally, future research efforts required to develop such safety specifications is identified.

1.4.3.1 Level I. Coupon and Element Level

The first level, or base, of the Building Block approach focuses on evaluating material behavior. For aerospace composite structures, this level typically focuses on coupon-level testing that is used to obtain quasi-static material properties (stiffness and strength), as well as to investigate notch sensitivity, fatigue resistance, and environmental effects. These properties are viewed as among the most important for preliminary design and analysis. For the design of composite structures for crashworthiness, however, additional coupon-level testing will be required for both material/laminate screening purposes as well as to determine crashworthiness-specific properties and parameters for use in computational analyses. Simple "element-level" testing may be required, wherein the geometry of the test article is intended to be "representative" of the intended application. Of the element-level test articles used to date for composite crashworthiness, untapered tubes of either square or circular cross section are most commonly used.

For aerospace composite structures, coupon-level testing to obtain stiffness and strength properties of a material is performed on the lamina level (using unidirectional composite laminates). For crashworthiness, however, it is important to consider the actual composite laminate proposed for design, since fiber orientations, stacking sequence, layer thicknesses, and total laminate thickness all may influence the energy absorption and general crush characteristics. Thus, Level I testing for crashworthiness will be required to be performed at the laminate level.

In addition to the characterization of the composite materials, additional Level I characterization is required to evaluate the usage of adhesively bonded connections for plastics and composites. Although considerable progress has been made towards the development of test methods to characterize adhesive bonds [29], less attention has been focused on the crashworthiness of bonded composite structures or strain rate effects in adhesives. Among the properties of greatest interest for crashworthiness are strength, fracture toughness, fatigue performance, and strain rate effects.

Proposed Safety Specifications and Metrics

At the material behavior level, proposed safety specifications will be associated with key crashworthiness-related properties of composite laminates intended for use in crush structures. A prescribed value of each property may serve as a metric for the associated safety specification. Several crashworthiness-related properties are described below from which safety specifications may be generated. A more complete discussion of crashworthiness properties, methods of measurement, and current status of such test methods used for testing is presented in Chapter 4.

Specific Energy Absorption (SEA): Defined as the energy absorbed per unit mass of crushed material. Although the SEA is currently the most recognized measure of the crashworthiness of a composite material/laminate, its usefulness typically is limited to material/laminate screening and ranking purposes. However, threshold values of SEA could be utilized for future PCIVs as a materials-related safety specification intended for composite crush structures.

Sustained Crush Stress: Defined as the average crush load divided by the specimen cross sectional area. Similar to the SEA, this property provides a measure of the crashworthiness of a composite material/laminate for use in screening and ranking purposes. However, the sustained crush stress is also useful in the design of crush structures. Threshold values of the sustained crush stress could be utilized in safety specifications of composite crush structures.

Compression Crush Ratio: Defined as the ratio of the compression strength to the sustained crush stress of a composite laminate. This ratio may be used as an indicator of the likelihood of the composite material crushing in a stable manner. As a result, this parameter is viewed as an important safety metric, with a threshold value being defined as a possible material safety specification.

Specific Static Strength: Defined for both tensile and compression independently. This attribute is important for the design of the passenger safety cell and effectively governs resistance to intrusion.

Laminate and Adhesive Fracture Toughness: Damage tolerance is an important aspect in all areas of the vehicle. It is inevitable that in some crash events, relatively minimal states of damage will be inflicted to the underlying structure. The energy required to propagate the failure through the composite material is important for analyzing possible failures of the safety cell.

Research Efforts Required

A significant milestone towards defining these material-level safety specifications is the development and standardization of suitable test methods. Although significant progress has been made in recent years towards the development of crashworthiness test methods, no standardized test method currently exist. As discussed in Chapter 5, further research is required to develop and standardize a flat-coupon test method for assessing the crashworthiness of composite materials. Additionally, further research is needed to develop and standardize an element-level tube test method for assessing composite crashworthiness.

1.4.3.2 Level II. Subcomponent and Component Level

The second level of the Building Block approach focuses on the design, development, analysis, and testing of sub-component and component-level structures. This level of the Building Block serves as a bridge between the base-level material-response determinations and the top level full-scale testing and analysis tasks. The distinctions between sub-components and components are illustrated in Figure 1-2 for the case of a complex cone structure. The component level complex cone retains the section properties of the proposed automotive structure and includes an adhesive bond and the complex features in the back end. In contrast, the sub-component level "plain cone' has the same cross section as the front end of the complex cone. In general, subcomponents used to assess crashworthiness are envisioned as being realistic, both in terms of size and shape, containing key features representative of the automotive component, but utilizing a simplified geometry. Such structures would be used for demonstrating crush characteristics, structural integrity of the back-up structure, and resistance to intrusion. Although there may not be any standardized subcomponents, it is believed that safety specifications may be developed for general categories of subcomponents that will aid in future PCIVs satisfying the required safety standards. The choice of both sub-components and components for testing and computational simulation will be highly dependent on the portion of the vehicle for which the article is intended (ex: front crush member, door structure, roof, floor panel).

The complex cone component illustrated in Figure 1-2 is representative of the ACC Focal Project 3 upper longitudinal structure, located forward of the A-pillar to behind the headlight area. Features in the component would be typical of those produced due to packaging considerations. The progressive crush in the front of the structure is disrupted by the local depressions which would typically be introduced late in a design to give local clearance to other components. The ability of the component to remain stable in a crash is further compromised by two features in the back-up structure: the large aperture (representing a strut mounting) and the swan neck which provides notional clearance for the powertrain installation.

For future PCIVs, this intermediate level of the building block is expected to include a focus on understanding the crush characteristics of composite structures through both testing and analyses. Of particular interest is the development of composite structural components that exhibit a stable crush front without failure in the backup structure.

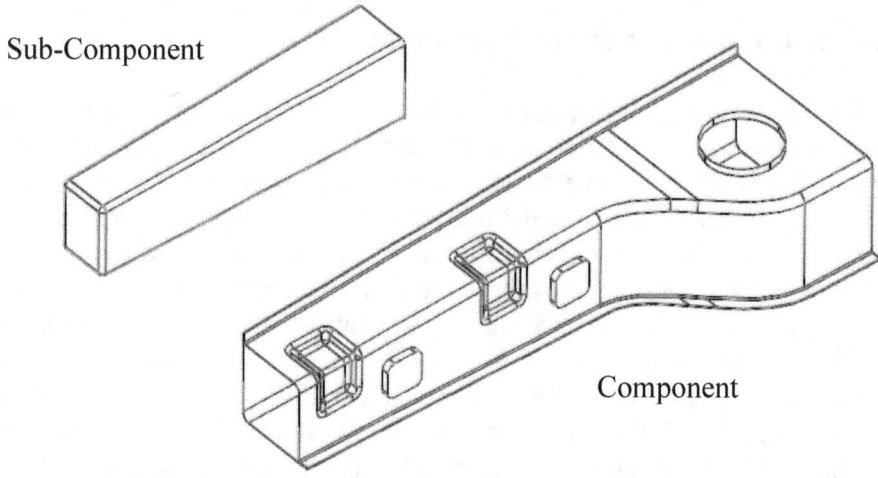

Figure 1-2. Sub-component and component level cone structures.

Computational analysis will play an important role in this level of the Building Block. In fact, computational analysis is typically utilized in the design of subcomponents for subsequent testing. In general for composite structures, emphasis at this level is correctly simulating the correct failure locations and failure modes due to prescribed loading scenarios. For crashworthiness, an added requirement is correctly predicting the crush response and the subsequent energy absorption during a prescribed crash event.

Following the design and manufacturing of the subcomponents or component, actual testing will be performed. The testing can serve two different purposes: to validate the computational simulations and to demonstrate the performance of the component under a critical loading condition. For the latter purpose, it is important that the test articles be manufactured to be as representative as possible to the production parts, both in materials and manufacturing process. For crashworthiness, the testing of substructures should be robust and inclusive of the various types of loadings produced from crashes in which the component is intended to provide energy absorption as well as resistance to intrusion.

Proposed Safety Specifications, Milestones, and Metrics

At the subcomponent and component level, proposed safety specifications may be developed for both analysis (computational simulation) and for testing. For analysis, a safety specification could be based on the accuracy of the computational modelling approach at predicting the force versus displacement response and failure mode(s) during testing in the preferred (as designed) directions as well as alternative directions which induce complementary failure and damage evolution. While defining a quantitative metric associated with this specification may be difficult, the intent is to ensure that the simulation predicts the general crush sequence observed during testing and therefore provides confidence in the ability to predict performance in scenarios not subsequently tested. For subcomponent and component crush testing, possible safety specifications could be based on the specific energy absorption or sustained crush stress produced. Outcomes of such a test could include demonstrated integrity, local damage, or structural collapse. Additionally, the performance of composite components can be readily compared to the "baseline" performance of the equivalent metallic structures from a conventional vehicle, demonstrating "baseline" performance. Regardless of the outcome, the computational simulation should correctly predict the general structural response.

Research Efforts Required

As will be discussed in more detail in Chapter 5, further research is required to develop current modeling approaches for predicting the crush behavior of composite structures. Composite subcomponents or components utilized in Level II of the Building Block can be used for benchmarking modeling approaches since they will be supported with testing.

1.4.3.3 Level III. Sub-Assembly Level

The third level of the Building Block approach is composed of multiple components as well as the added complexity of their assembly and interaction. The size and complexity of the sub-assembly is dependent on the associated function within the vehicle. For example, a sill and floor could be useful to investigate the crashworthiness associated with a side impact. In other situations, however, a much larger sub-assembly may be required, such as the entire passenger safety cell without the surrounding energy absorbing structures.

As a second example, consider the connection at the top of the A-pillar with the header, cant rails and roof. Added complexities due to adhesive bonds and mechanical connections will be incorporated. This joint is critical to the integrity of the roof during crush loading, but will also be subjected to considerable loads in a frontal crash and potentially in a side impact as well. This sub-assembly, composed of multiple components, will require testing and analysis under multiple loadings. For testing, the subassembly would need to be extended to rigid supports, loaded, and the failure mechanisms recorded for comparison with analysis predictions.

Proposed Safety Specifications, Milestones, and Metrics

The sub-assembly level represents a major milestone on the route to compliance with the FMVSS for roof crush, but also will be evolved to be a key metric in the specification of passenger safety cell integrity. Furthermore, sub-assembly testing will be comparable with conventional steel structures in mass usage today. The performance of sub-assemblies of composite components in impact tests will undoubtedly result in different failure modes compared with the collapse of a steel structure. A useful performance metric may be the load at catastrophic failure or at which the associated safety cell decelerations would be non-survivable by the occupants. Using a load/survival space ratio, this maximum load may be compared with conventional vehicle structures in use today.

Research Efforts Required

In addition to research efforts from previous levels, joining technology will be extended to test production intent joint configurations at this stage. Adhesives and mechanical attachments that have the necessary durability, strength and ruggedness will need to be developed, through analysis and test. This development must also include practical analysis tools necessary to allow the prediction and hence development of assembled parts in impact related load cases.

1.4.3.4 Level IV. Full-Scale Level

The top level of the Building Block includes analysis and testing of a full-scale structure. For the case of crashworthiness assessment of future PCIVs, this level will focus on crash simulations and tests involving the entire PCIV. Both are expected to focus on establishing compliance with Federal Motor Vehicle Safety Standards (FMVSS).

It is expected that considerable full-scale vehicle simulations will be performed during the design and development stages of future PCIVs such that full-scale vehicle testing serves to validate the computational modelling as well as ensure compliance with FMVSS. It is expected that the same loading cases (those specified by the FMVSS) will be utilized for both cases to minimize full-scale test costs. Due to the test costs associated with full-scale vehicle testing, it is expected that any additional tests required for model validation will be performed at the component level.

Proposed Safety Specifications, Milestones, and Metrics

Since full-scale testing is expected to address governing safety standards directly, it is not expected that any specifications will be required for such full-scale testing. However, it is possible that a future safety specification could be developed for the computational simulation of a full-scale vehicle. As for component analysis, such a safety specification could be based on the accuracy of the computational modelling approach at predicting the force versus displacement response during testing as well as the observed crush sequence.

Research Efforts Required

Similar to Level II, further research is required to develop computational modeling approaches for predicting the crush behavior of PCIVs. Full-scale PCIV crash testing may be used for benchmarking candidate modeling methodologies.

2. LESSONS LEARNED FROM COMPOSITES IN HIGH PERFORMANCE CAR APPLICATIONS

2.1 Introduction

Development of mass market Plastic and Composite Intensive Vehicles (PCIVs) has been an aspiration for the automotive industry for several decades. To date, however, only limited production, high-performance supercars have extensively used composite materials. This chapter summarizes findings obtained from both the racing industry as well as from high-end, limited-production commercial automobiles. Additionally, lessons learned from other published research activities pertinent to PCIV development are summarized.

2.2 Lessons Learned

Automotive applications currently represent one of the top market potentials for composites. While structural applications in vehicles continues to be viewed as a significant market opportunity for composites utilization, there has been resistance by the major car manufactures to develop structural applications for composites in the volume segments of the market [30]. As a result, currently there is limited information to justify the safety of composite-intensive structures based on normal road use and typical crash scenarios. Attempts to utilize safety information from racecars or supercars with composite structures as a barometer for safety is distorted by the nature of their speed and the corresponding damage induced. However, there is value in reviewing information relating to the adoption of composites into Formula 1 car structures, and their more intensive application to the Le Mans cars as well as the more extensive "real world" crash portfolio to have affected the modern composite-intensive $500k+ supercars. This review is intended to identify attributes of the composite structures and their performance in crashes that can be used as reference points for future development of mainstream high volume PCIVs.

2.2.1 Formula 1

One of the first major usages of composites in Formula 1 (F1) racing was the development of a composite chassis "safety cell" in 1980 by McLaren in the form of the John Barnard designed MP4/1. The composite monocoque chassis was molded around a complex aluminium multi-part mandrel which was removed through the cockpit aperture after the three-stage curing process. This chassis was predominately a carbon/epoxy composite chassis with aluminum honeycomb as a core material [31].

Considerably concern was voiced over the ability of these "brittle" materials to survive impact, and at the time no substantial data existed. Inevitably, crashes occurred and the initial fears of the structures "shattering" were dispelled. In some relatively major crashes, the damage to the composite parts of the structure was localized and easily repaired with no apparent degradation in performance. According to the February 2006 edition of *Racing Line*, the McLaren Group's in-house magazine, the driver of the McLaren MP4-1 car, John Watson stated:

> *"A composite carbon fiber chassis was a big step into the unknown," he says. "The question all Formula 1 drivers were asking was what was going to happen in an accident?" The Ulsterman found out early in the test program that the unyielding nature of the carbon fiber was very different to the steel and aluminum panels he was used to. Team-mate Andrea de Cesaris demonstrated the car's structural integrity by walking away from a number of crashes. Watson found out for himself when he escaped from a 140mph crash that destroyed the car at Monza's daunting Lesmo bends. "Fortunately, the design turned out to be virtually bulletproof," he says. "It's easy to take a new material and apply old thinking, and many people didn't understand the technology at first. But John Barnard and his team weren't into gambles - they knew exactly what the materials would give them. The MP4-1 was born out of incredible vision. [32]"*

It was not until 1985 that the first frontal crash tests were introduced, and by this time all teams were utilizing the carbon fiber reinforced composite monocoques, with the majority molded in female tools and in many cases joined along a constant z-plane through a tongue and groove or banged glue joint. Since that time, crash test requirements have increased and now include preconditioning tests to ensure that the nose cone of the race car remains attached following a minor oblique impacts in case of a follow-on axial impact. Before the nose cone and monocoque structures are tested for the front crash case, the nose cone push-off loads equivalent to 40 kN (9.0 kips) is applied laterally to the side of the nose cone 550 mm (12 in.) in front of the wheel centerline. Similar tests are applied to the side impact tubes and the rear impact structures before testing [33].

As the first generation of composite chassis designs progressed, based on mainstream commercially available carbon/epoxy composite materials, the competition for increased performance sent designers and materials specialists in search of stronger and stiffer materials in order to further reduce the weight of the structure. However, the single regulatory test at the time for front impact potentially masked the need for the composites to be robust in many different crash scenarios. Ironically, a nose cone test places a relatively even load distribution into the safety cell that does not occur in various loading scenarios associated with crashes.

Several crashes involving Formula 1 car structures have demonstrated the safety attributes of composite structures. In September 1990, a crash occurred at Jerez, Spain involving an F1 Lotus driven by Martin Donnelly. The car impacted the Armco almost perpendicularly at a speed estimated to be approximately 225 km/hr (140 mph), and the driver survived, although with significant injuries. The front of the Lotus 102 chassis was observed to literally disintegrated between the front axle to the fuel tank bulkhead. Following the crash, the driver remained attached to the GRP seat in the middle of the track. Although little information is published on the technical aspects of this crash, it is fair to conclude that this was not the expected mode of failure in the chassis. Undoubtedly the energy was absorbed in the barrier and in the fragmentation process, but today it would be expected that the safety cell of the monocoque would be left intact.

In 1994, static tests were introduced on the chassis structures to improve their resistance to failure of the safety cell. The following year, the Federation Internationale de l'Automobile (FIA) introduced a side impact test to the chassis to ensure the designs would absorb energy in the outer footprint of the vehicle. In 1997, rear impact structures were introduced on the back of the "stiff" gearbox/power train. Side intrusion panels were specified in 2001, which extended the side coverage of the driver. New tests were defined to ensure that these panels offered sufficient resistance to nose cone impact from another vehicle or foreign body penetration.

In general, the introduction of new safety tests and protective structures helped ensure that safety was not compromised while focusing on weight reduction and maximizing torsional stiffness. Such safety measures also promoted the use and accelerated the development of toughened resin systems and fiber combinations to absorb energy and produce local failure. As a result, the fatality rate in Formula 1 plummeted from 1 in 40 crashes in 1980 to 1 in 250 crashes in the following 12 years. Since 1994, no fatalities have been recorded in the category, despite some horrific high speed impacts which in the preceding years would likely have resulted in fatalities. Through this period, the fundamental open-wheeled design, the location of the driver, and the use of multipoint harnesses and helmets remained unchanged.

It is important to recognize that in addition to the inherently improved safety structures of the chassis, there were other measures invoked for the circuit design and additional restraint systems (HANS device) that also greatly contributed to these safety improvements. The Formula 1 teams strive to meet the impact regulations while minimizing weight of the chassis and optimizing stiffness and aerodynamics. Unlike production cars, Formula 1 cars are not rated for crash safety. Rather, the performance standard is a pass/fail test, based on the ability to absorb energy without exceeding limiting acceleration levels.

Considering the multitude of possible crash scenarios for a Formula 1 car, the energy absorbed by the system and the maintenance of structural integrity is impressive. This behavior can largely be attributed to the use of composites in the chassis. One particular advantage is the ability for the composite structures to absorb energy locally at the site of impact. If the vehicle continues to suffer multiple impacts, further energy may be absorbed either locally at the site of impact or elsewhere in the vehicle. Additionally, the vehicles often may be repaired after impacts and returned to a fully functioning condition. As an example, a Honda F1 went on to claim the team's first ever victory at the 2006 Hungarian Grand Prix following extensive repair to the front portion of the monocoque chassis [34].

Although the repair procedures are significantly different from those employed for conventional vehicles, composite structures are able to be repaired to a level of performance comparable to a repaired metal structure. Such repairs have been achieved in the racing car industry at racing levels below that of Formula 1, where crashes are more frequent. At such levels, repairs to the composite chassis are typically made by the teams running the cars rather than the manufacturer.

2.2.2 Le Mans

In the late 1990s, the Le Mans 24 Hour Endurance Event's Premium Class moved towards road-legal cars which had passed a "rigid frontal crash test". The rules were designed to make the race cars derivatives of high performance road cars, but the initial effect was to make cutting-edge race cars that could operate on the road. As a result some significant composite racing cars were evolved with frontal crash testing a requirement.

The McLaren F1 was already in production: a total of 106 units were produced in road and race variants. The chassis had passed frontal impact testing and the car was being used for the Le Mans competition with considerable success. This car was the original baseline for other manufacturers such as Mercedes, Panoz, and Porsche.

AMG, on behalf of Mercedes, produced a series of three models starting in 1997 with the aim of contesting the FIA GT championships. They performed crash tests with the car in time to allow further derivatives for Le Mans the following year. The 1997 FIA GT car was exceptionally successful on the track, and resulted in the team winning the Drivers and Constructors Championship. However, there were difficulties associated with meeting the crash test requirements as the loads transferred by the nose cone were not adequately reacted by the safety cell. Mercedes contracted specialists to develop the chassis. The changes required to meet and comprehensively exceed the crash requirements were modest and easily retrofitable to the series of chassis that had been produced for the road.

The design of the 1998 car had already commenced with a much closer attention to reacting all the required crash loads and developing the performance aspects in conjunction with the safety structures. The design team was interested in pursuing an all-composite roof structure, and yet there was some reluctance based on issues with the previous car. A series of impact tests were performed on a contemporary welded roll-cage structure from a German touring car, and a corresponding composite structure of the roof. Whereas the welded structure failed on impact at the "brittle" joints, the composite roof structure was able to withstand the impact with localized damage at the impact point. Furthermore, this structure was able to repeat the exercise a number of times with progressively increasing damage at the impact point but without compromising the survival space. The decision was initially taken to utilize composites but to increase the design requirement by 100% and validate it on an additional chassis to prove confidence in the structure. Later, however, the decision was changed to also include a lightweight metallic roll structure in parallel.

The 1999 assault by Mercedes on Le Mans was more significant. The vehicle was designed and constructed with a composite roof and a composite chassis structure, which was understood to be a first for Le Mans. The performance benefits of a lightweight roof with less obscuration from the A-pillar were desirable. The chassis and roof structure were designed and tested against the FIA's compound loadset which gives (Vector) at approximately 89 kN (10 tons).

The Mercedes was flawed by an aerodynamic imbalance that caused the front of the vehicle to lift un-recoverably and take-off at speeds approaching 320 km/hr (200 mph). Unfortunately this was also the case with a number of competitor's vehicles and was considered to be aggravated by the rules concerning the flat floor. In a series of incidences over the course of the event, the car crashed heavily on these structures a number of times. In the first incident, flipped end-over-end, and landed hard on the rear wheels. The chassis was damaged, but the driver walked away unhurt from the incident. No photographs are known to exist of the crash. The chassis was replaced and the same driver and car essentially repeated the feat two days later. This time, the car landed on its roof from a considerable height, absorbing energy locally in the roof structure. However, the roof maintained structural integrity with no reduction in survival space. The driver again walked away from the car unscathed. Photographs of this incident are available at [35].

The third incident is the most shocking and most graphically represents the magnitude of the events. The vehicle leaves the track airborne at close to 320 km/hr (200 mph) and lands off of the track and into a conifer tree with a trunk of approximately 250 mm (10 in.). diameter. The impact crushed an estimated 300 mm (12 in.) of material in the rocker and the cant rail. Most importantly, while dissipating considerable energy, this crushing did not violate the integrity of the safety cell. The chassis was propelled approximately 30 meters (100 ft) to its final position. The safety cell was intact and the driver stumbled from the wreckage, was checked out, and given the "all clear" at the circuit. This crash was caught on video and is available on-line for viewing at [36].

The Le Mans Prototype (LMP) category was afflicted by the same aerodynamic issues that affected the Mercedes. The web-based article "When Le Mans Racecars Fly" on the Popular Science website [37] provides additional examples of the immense strength of the chassis and the ability of the cars to absorb energy. The Porsche 911 GT1 from Road Atlanta performed an impressive back flip, landing on it's rear structure followed by a 160+ km/hr (100+ mph) oblique impact into the side wall. There was extensive destruction of all energy absorption devices while the safety cell remained intact, protecting the driver. A more recent example involved a high speed oblique side impact of the Peugeot 2008 HDI into a solid wall, with the side structure remaining intact. An interesting insight in to the variety of impacts on these cars is illustrated by the Courage-Oreca LC70 crash in 2008 at Monza. The car is pitched into a series of end-to-end cart wheels at high speed before a final impact to the base of the chassis against the side wall. The integrity of the chassis is all the more incredible due to the early loss of the nose cone in the incident as shown in the photograph at [38]. Despite this, the driver escaped with a minor fracture of the lower leg.

An insight into the developments of the Le Mans Prototype (LMP) cars, particularly the closed car series developed by Bentley, is discussed in [39]. The different structure for managing roof crush for all-composite designs led to a significant change in the structure that had been used before.

In summary it is important to note that crashes of the magnitude presented above, which have been proven to be survivable, are not to be regulated for on the road. However, the ability of these racecar composite structures to survive huge impacts and absorb the energy through local failure is an indicator of the benefits obtainable through using composites in the energy absorbing structures and passenger safety compartments of future PCIVs. The chassis safety cells of the racecars discussed above are approximately 70 kg (154 lb.) in weight and optimized for the task of safety and rigidity. There will be a reduction in inherent "geometric" safety as the structures are translated into mainstream vehicles, since door sizes will increase and structural rocker, headers and A-pillars will necessary be reduced.

2.2.3 High-End Supercars with Composite Safety Cells

The McLaren F1 preceded the Le Mans cars discussed above. This car pioneered the use of carbon composites for road use, although only 65 road versions of these $1 million cars were sold. With a significant number of the vehicles in the hands of museums and private collectors, fleet mileage and crash data is limited. During the development of the vehicle in hot weather testing in the Namibian desert, a test driver hit a rock and the car rolled a number of times. However the driver emerged unharmed. The car did however meet the UK frontal impact regulations and was tested in late 1994. As seen in an online video [40], the nose cone absorbs the impact energy with very little deflection being observed in the safety cell.

The Le Mans composite GTS car was the impetus that lead to the development of three road supercars. Although they never raced at the highest level at Le Mans, they were produced in reasonable volumes: 1,270 Porsche Carrera GTs, 400 Ferrari Enzos and 50 of its sister car, the Maserati MC12, and 200 Bugatti Veyrons. While crash data for these vehicles is limited, the Carrera GT and Enzo have been involved in a number of crashes. Although these crashes have being relatively undocumented from a scientific perspective, photographs of the aftermath do yield some interesting findings on the performance of the composite chassis structures in impact. It is also worth keeping in mind that the crashes that these cars are involved in typically involve excessive speed with corresponding increases in energy over "mainstream" automotive crashes.

The Porsche Carrera GT safety cell is an all-composite structure that includes removable roof panels, placing greater emphasis on the strength of the rockers. A series of crashes have been documented involving the Carrera GT which address the crashworthiness of the composite safety cell. The following is a brief summary of several of these crashes.

- A series of flips to a car from high speed in Portugal caused extensive damage to the front and rear structures of the vehicle, but the passenger safety compartment remained intact, with the doors still operating as shown in Figure 2-1.

Figure 2-1. Porsche Carrera GT following flipping over several times [41] (used with permission).

- A severe pole impact to a Carrera GT produced no apparent intrusion or collapse of the safety cell, despite extensive crush and destruction for the front end as shown in Figures 2-2 and 2-3.

Figure 2-2. Porsche Carrera GT following pole strike [42] (used with permission).

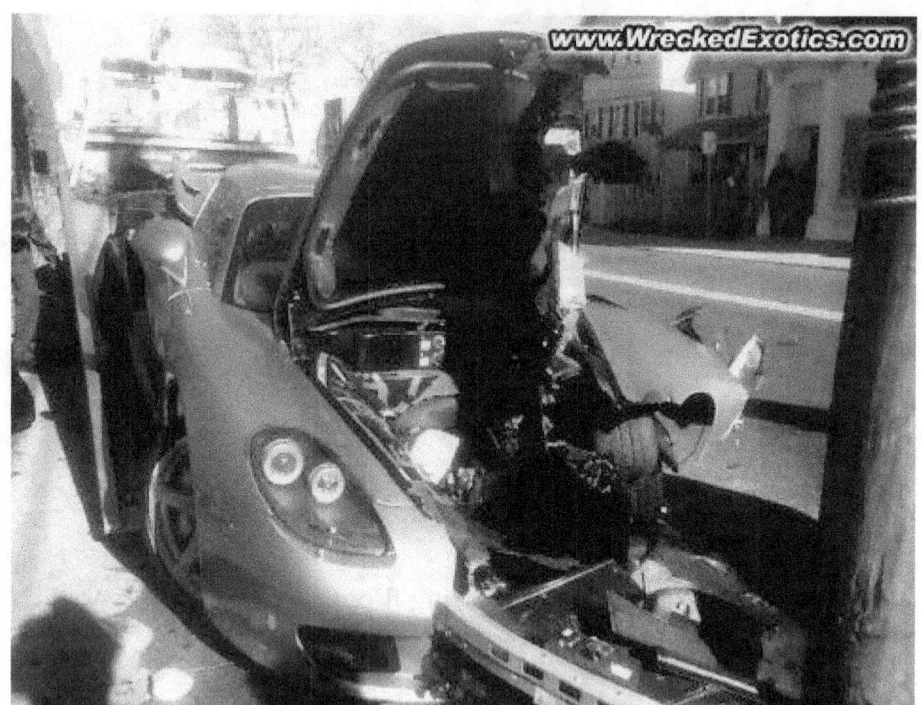

Figure 2-3. Front view of Porsche Carrera GT following pole strike [43] (used with permission).

- A side impact of a Carrera GT into a tree injured the passenger seated on the opposite side to the impact (five broken ribs). However the damage to the cabin was minimal and localized at the point of impact, where the composite structure experienced localized crushing and absorbed the energy as shown in Figure 2-4.

Figure 2-4. Porsche Carrera GT following side impact [44] (used with permission).

- A crash at a racetrack which resulted in the death of the driver and passenger in a Carrera GT at the California Speedway. The crash involved the car leaving the track and impacting a concrete barrier at approximately 240 km/hr (150 mph), substantially sideways. The safety cell is apparently intact and the non impacted door is shown to be operational as shown in Figure 2-5.

Figure 2-5. Porsche Carrera GT following side impact [45] (used with permission).

- A crash involving a side impact of a Carrera GT into a tree with the impact behind the cabin, causing the engine and gearbox to detach from the safety cell. This initial impact was followed by a frontal impact with another tree. The safety cell appears to be intact and both doors fully operational as shown in Figure 2-6.

Figure 2-6. Porsche Carrera GT following crash into tree [46] (used with permission).

An illustration of a crash on a much earlier supercar, the Bugatti EB110 (produced 1990 to 95, with 167 sold) indicated that a significant impact with a steel lamp post did not damage the safety compartment as shown in Figure 2-7. Although equipped with a composite chassis, the design of this vehicle is far less extreme than current era supercars. The rocker and cant rail sections are more appropriate for the design of a future PCIV.

Figure 2-7. Bugatti EB110 following side impact [47] (used with permission).

The chassis of the Ferrari Enzo is entirely composite, constructed using composite sandwich panels with carbon/epoxy facesheets and aluminum honeycomb core. A series of crashes have been documented involving these supercars which address the crashworthiness of their composite chasses. The following is a brief summary of three of these crashes.

- A side impact of a Ferrari Enzo into a tree at excessive speed resulted in the engine and power train behind the bulkhead being detached as shown in Figure 2-8. Fuel integrity was compromised, resulting in a significant fire. The impact point appears to have been just behind the chassis joint, which is likely to have caused failure of the fuel connections. However the passenger safety cell is intact after the impact, (see Figure 2.9), indicating that without the fire and despite the high speed impact directly on the side of the vehicle, the occupants may have otherwise escaped serious injury.

Figure 2-8. Ferrari Enzo following side impact [48] (used with permission).

LESSONS LEARNED FROM HIGH PERFORMANCE CAR APPLICATIONS

Figure 2-9. Ferrari Enzo "loosely re-assembled" after side impact [49] (used with permission).

- A significant front impact of a Ferrari Enzo produced approximately $400,000 in damage, but an apparent lack of damage in the passenger safety cell, see Figure 2-10. All closures will still functioning following the crash as seen in Figure 2-11.

Figure 2-10. Ferrari Enzo following frontal impact [50] (used with permission).

Figure 2-11. Ferrari Enzo following frontal impact. Note functioning closures [51] (used with permission).

- Another high speed crash involving multiple flips and rollovers of a Ferrari Enzo resulted in the engine being detached from the chassis. Despite the safety cell being modified to have larger roof apertures, it remained substantially intact and the survival space was maintained as shown in Figures 2-12 and 2-13. The impact lead to broken vertebrae and ribs in the driver.

Figure 2-12. Ferrari Enzo following multiple flips and rollovers [52] (used with permission).

Figure 2-13. Ferrari Enzo, view from opposite side [53] (used with permission).

The Mercedes-Benz McLaren SLR is equipped with an all-composite safety cell as well as a composite front crush structure which have been documented as providing excellent crashworthiness in crashes. The following is a brief summary of two such crashes.

- A McLaren SLR underwent multiple rolls as a result of a high speed crash in the Qatar desert. Although the crash resulted in a double fatality, the safety cell remained intact. The wreckage from the portion of the vehicle in front of the passenger cabin was completely destroyed as shown in Figure 2-14.

Figure 2-14. McLaren SLR following multiple rolls from high speed crash [54] (used with permission).

- The integrity of the McLaren SLR safety cell was demonstrated by a high-speed offset frontal collision with a Volkswagen Golf. The three occupants of the SLR were able to walk away from an intact safety cell while the driver of the Golf was left with serious injuries and multiple fractures. A Photograph of the SLR following the collision is shown in Figure 2-15.

Figure 2-15. McLaren SLR following high speed front offset impact [55] (used with permission).

There are limits to the performance of any structure, including the composite safety cells of high-end supercars. A single-car, 240 km/hr (150 mph) crash in Milan, Italy involving a Ferrari Enzo demonstrates what can happen when those limits are exceeded. In this case the crash resulted in a number of impacts, and the vehicle structure was torn apart and the safety completely violated, see Figure 2-16. While the aftermath is shocking and the crash debris quite different from what would be expected from an aluminum or steel supercar, the results from a survivability standpoint are expected to be the same. While the load limits of a composite safety cell may be considerably higher than the those at which aluminum or steel cabins would collapse, there is no hiding from the fact that the failure mechanism of the composite safety structure at the limit involves brittle fracturing. Loaded above their maximum load level, such composite safety cells will "disintegrate." It is important that all stakeholders understand that this kind of catastrophic destruction is expected and not a fault of the design of the vehicle.

Figure 2-16. Ferrari Enzo following 240 km/hr (150 mph) crash [56] (used with permission).

It is apparent when comparing crashes involving supercars with composite safety cells with similar crashes involving aluminum or steel structured vehicles that the composite safety cell generally provides greater integrity, often to a level above which the occupants can tolerate the acceleration levels without serious injury. The metallic cars generally exhibit considerable intrusion before such acceleration levels. Such intrusions are often the primary cause of injury, Figure 2-17 and 2-18 show the aftermath of crashes involving steel cars, which were involved in similar speed crashes to those that resulted in Figure 2-16 above, and resulted in significant intrusion and also partition.

LESSONS LEARNED FROM HIGH PERFORMANCE CAR APPLICATIONS

Figure 2-17. Porsche 911 following high speed crash [57] (used with permission).

Figure 2-18. Ferrari 360 Modena following high speed crash [58] (used with permission).

In side-impact crash scenarios involving poles, both vehicle classifications can be prone to the detachment of the engine and gearbox assembly. However, in the extreme side impact event on a metallic vehicle, the intrusion dominates the injury, where a composite chassis may have partitioned, or in any event at the loads necessary to cause the intrusion damage shown in Figure 2-19 would be expected to cause fatal injury due accelerations to the occupants in a composite chassis.

Figure 2-19. Porsche Boxster following extreme side intrusion [59] (used with permission).

2.2.4 Research and Development Activities on Composite Vehicles

The limited amount of published research activities suggests that PCIV development has been somewhat limited to date. However, several research and development projects have been performed in recent years. One of the more significant activities has been the Automotive Composite Consortium, part of the United States Council for Automotive Research (USCAR) [60]. This partnership between the U.S. Department of Energy and industry (Ford, General Motors, and Chrysler), focuses on *"joint research programs on structural and semi-structural polymer composites in pre-competitive areas that leverage existing resources and enhance competitiveness* [61]". The ACC has funded a variety of research projects related to crashworthiness of composite materials. Additionally, the ACC has sponsored "Focal Projects" to develop and demonstrate technologies. The latest to be completed, Focal Project III focused on designing, analyzing, and building a composite intensive Body-In-White (BIW) structure [62]. This study considered the crash safety implications of composite intensive structures. With a 67% saving over steel, the BIW design had considerable reserves in the safety cell under front crash conditions. Demonstrations of the front lower longitudinal were made at the time with a monolithic/wrapped or braided layup. Good Specific Energy Absorption (SEA) levels of 46 kJ/kg (15.4 x 10^3 ft-lb/lb) were obtained. Complex front upper longitudinal members were demonstrated to absorb the required energy levels of 3.6 kg (7.9 lb) per side using woven fabric/epoxy. Currently Focal Project IV is underway, focusing on the design, analysis, fabrication, and testing of a structural composite underbody as well as a second-row composite seat.

Composite materials utilizing chopped fiber have been utilized in the production of Aston Martin automotive structures. More recently the use of this type of composite material has been extended and improved upon by Bentley Motors [63]. Using a directed performing approach, nearly 90% of the fibers were aligned using a high-speed chopping process.

In Europe, Technologies for Carbon Fiber Reinforced Modular Automotive Structures (TECABS) project developed a composite floor pan for the VW Lupo vehicle with a projected 50% weight saving and production at 50 units/per day [64, 65]. Only a floor pan was considered in detail, and no side impact investigations or front crash assessments or recommendations were made. However, the research did lead to some interesting developments in delamination modeling [66].

Following the completion of the TECABS project, the European Super Light Car (SLC) project was initiated, with the goal to *"reduce weight in vehicle bodies through the economically feasible production of multi-material structures"* [67]. Despite the fact that most of the TECABS collaborators were involved in this project, the composite floor pan concept developed previously in the TECABS project was not incorporated. The final Body in White (BIW) consisted of 50% steel and only 4% plastics, producing a modest weight savings of 30%.

The Japanese New Energy and Industrial Technology Development Organization (NEDO) funded a five-year CFRP Automobile Body project starting in 2003. The objective of this project was to *"design an automotive body which exhibits 50% lighter and 1.5 times higher impact energy absorption capability in the full wrap collision test compared with a current steel body"* [68]. The project focused on thermoset fabric composites, however various other materials were evaluated as part of the project.

Lotus Engineering's Project Ecolite [69] has focused on the development of a thermoplastic composite front end structure. Evolving from composite crash systems previously developed by Lotus for high-end vehicles, this composite crash system is intended to be economical such that is suitable for higher volume applications.

2.3 Conclusions

The introduction of composites into the motor racing environment in the early 1980's has dramatically improved the likelihood of surviving excessive impacts, and has been an important contributor the drastically reduced fatality rate in the sport. The use of composite materials in the closed Le Mans cars was a stepping stone in development to the modern composite-intensive supercars. The comparison of severe impacts in conventional metallic safety cell cars and the modern composite counterparts illustrates the improvements in energy absorption possible and the associated increase in occupant safety. Despite the evolution of the automotive racing industry towards composite structures, however, progress to date towards the development of PCIVs has been very limited.

Carbon composite structures used in contemporary racing cars and the supercars of recent years are becoming more applicable and affordable to the mainstream market. The high specific energy absorption, strength, and stiffness of these materials will present similar opportunities for application in future PCIVs.

The low density of the materials (less than one-fourth of steel) allows larger sections at greater thickness, which inherently reduces deflection and stresses in the structures and prevents intrusion. Lightweight composite passenger safety cells can remain intact and prevent intrusions far beyond the onset of collapse in a *comparable/conventional* steel structure. Furthermore, the forces necessary to trigger the collapse of a composite safety cell typically generate accelerations above which the occupant is expected to survive. It is important to ensure that the onset of local failure is prevented, and where unavoidable demonstrated not to initiate a catastrophic collapse through careful selection of materials and connections between the panels of the body structure.

The racecar industry has embraced the abilities of composites to absorb significant energy per weight, and composites are used regularly in front, rear, side, and roof crush structures. Racing officials have recognized the need to ensure that the energy absorption potential is realized in a crash scenario and have developed regulations to "test" the stability of the structure by first applying loads perpendicularly to the direction of impact to ensure that the structure is capable of surviving an oblique strike followed by a direct impact.

Composite safety structures are readily repairable. Techniques for assessing the extent of non-visible damage have been pioneered in other industries and are regularly in use in the racecar industry today. Extensive repairs are possible, and the practice of alignment and jigging applied to metallic structures will be necessary for major repairs to composite assemblies.

3. DEVELOPMENT OF STANDARDIZED TEST AND EVALUATION PROCEDURES

3.1 Introduction

The Plastic and Composite Intensive Vehicle (PCIV) has the potential to revolutionize the automotive sector, due to the inherent benefits of composite materials. These materials exhibit high strength-to-weight and stiffness-to-weight ratios as well as excellent energy absorbing capability per mass. However, the behavior of these composites in the crash and safety environment requires an in-depth knowledge of the materials and their unique performance characteristics in order for the designer to get the best from these new materials. For several generations, automotive designers have used steel for structural design. The current and future generations of automotive designers will need to be educated in the best use of composite materials for energy-absorbing automotive structures. Unfortunately, redesigns based solely on material substitution from steel to composite will not provide the realizable benefits that composite materials are capable of. Future automotive designers will need to develop new designs based on the new materials and utilize new and yet-to-be-developed standards and material databases for composite materials.

The behavior of most composites under large deformation load cases is radically different than the plastic deformations produced in metallic automotive components. As such, their behavior must be well understood by the designer to fully utilize their benefits while not being hampered by their characteristic brittle failure modes. Steel construction allows the designer great freedom since the ductility of steel allows for partial failures away from the main energy absorbing structure without compromising the overall structure. Composites, however, are somewhat less tolerant to such partial failure behind the region of crushing. Additionally, composites absorb energy by fragmentation of the material, which leads to the destruction of the part. In order to achieve high levels of energy absorption, the crush progression of the composite must proceed in an appropriate order. Crushing must initiate at the front of the structure and proceed through the structure like a wave front. If a failure is produced away from the crash front during this process, the structure will lose considerable energy absorbing capability. Thus, the structure must have sufficient strength in the back-up structure (behind the crush front) to fully support the forces being generated. In this respect composite crash structures tend to be less forgiving than their metallic equivalents which benefit from the ability of the material to undergo much larger plastic strains before rupture or tearing.

The design of future PCIVs will require a progression of steps that will include material selection/evaluation, preliminary design, concept stages, tests and evaluations. A concept stage will be required, where a range of possible solutions are considered that appear to satisfy the requirements of the PCIV structure. These concepts then need to be evaluated to identify the most promising design(s) to pursue. The choice of materials for the vehicle structure is crucial in the concept phase, as different fibers and architectures produce different energy absorption characteristics. A database of properties will be required for material selection, or at a minimum a set of standard test procedures such that data may be reliably and consistently obtained. Having a concept and a list of potential materials, the designer must develop the design to a working concept, meeting all the regulated tests and any additional in-house requirements. This will require effective, correlated analysis tools that are capable of simulating the crush and failure mechanisms associated with composite materials. Such analysis tools are currently being developed, and are at various stages of development. To ensure that these design concepts are achieving their anticipated performance criteria, a well-developed test procedure should be followed.

In this chapter, changes and additions to test and evaluation procedures due to PCIVs are discussed, with a focus on ensuring their compliance with Federal Motor Vehicle Safety Standards (FMVSS)

3.1.1 Candidate materials

In the definition of PCIV proposed in Section 1.2, there is an emphasis on the structural attributes of the material to react load and prevent intrusion, but also to be able to absorb energy in an impact. There are many plastics and composites incorporated into current vehicles for other reasons which are not safety related and not discussed in this document.

Reinforcements are perhaps the easiest of the candidate materials to classify. While most engineers will pick carbon fiber for the reinforcement based on performance, the costs of carbon still remains a commercial barrier. However, results from pilot production of a low cost production process pioneered at ORNL shows promise [70]. From a weight efficiency standpoint, carbon fiber provides the best stiffness, strength and energy absorption performance. Thus it is not surprising it is highest on the list of composite reinforcements, and also the target of considerable research to reduce the costs to commercially- viable, high-volume levels. Further commitment to the selection of carbon as the viable fiber has been given by the decision by SGL/BMW to commence the construction of a carbon fiber plant at Moses Lake, Washington, to satisfy the large scale production of their MegaCity Car [71].

Alternative reinforcements include glass, aramid, Ultra High Molecular Weight Polyethylene (UHMWP) and natural fibers such as flax, hemp and sisal. While aramid and UHMWP typically show higher tensile strength performance than carbon, their lack of ability to bond to themselves and other fibers reduces their compressive performance. Little work has been reported on long fiber applications on the natural fibers, particularly with respect to energy absorption. However, they show good characteristics for bonding the fiber bundles. Their inherent surface finish may promote increased interlaminar and through-thickness performance, which may be beneficial for crush stress. Further research should investigate the opportunity of using these low cost, low density fibers in these structural applications.

The wide variety of polymer-based plastics which are used as matrix material for fiber composites are too numerous to list. Matrix materials suitable for high volume will need to be processed quickly. Candidates will include both thermoplastic and thermoset materials. Much of the supercar experience has been gained using epoxy based thermoset systems used in prepreg laminates. Lower-cost and shorter processing time versions are being developed for infusion processing. The plastic materials supply chain needs to recommend cost effective, high production materials which can be readily paired to a wide range of carbon and alternative fiber types.

Core materials, such as the established family of closed-cell foams, have become widely incorporated within vehicle structures and interiors to improve the safety performance. In applications where they can reinforce members for either axial crush or out-of-plane impact, they work to support the primary structural material which could be based on conventional metallic or advanced PCIV construction. Lightweight core materials do not significantly affect the PCIV classifications as their low density and weight do not contribute highly to the percentage plastics and composites composition.

Further research effort should be channeled to assessing the key functional attributes of core materials for structural reinforcement and energy absorption. Similarly, lower cost higher performance plastic derivatives should be developed. Progress in this activity will result in benefits feeding into the vehicle safety design through greater efficiency.

3.2 Overview of Federal Motor Vehicle Safety Standards Relevant to the Development of PCIVs

The current regulations addressing safety standards for conventional steel vehicles has been developed over a number of years in response to the need to reduce the frequency of injuries associated with automotive crashes. Additionally, the accumulation of crash data has been useful for determining those crash scenarios that are likely to result in injury and in designing tests that can be used to improve the vehicle structure in such crashes. When considering the test requirements for PCIVs, the inherent lack of ductility in composite structures may require changes to existing test procedures to ensure that small variations in the test set up do not lead to dramatically reduced energy absorption. As an example, roof crush tests performed on a steel vehicle assess the ability to withstand a force applied through a flat platen at a specified angle. Resulting roof structures are comprised of substantial sections in the A-pillar, header and cantrails. Due to its ductility, this roof structure is expected to be robust in also resisting variations of the roof crush test (shaped platen, different angle, etc.) In contrast, a composite roof structure designed for the specific test arrangement may be vulnerable to the sharper impact or having the load applied in a different manner (for example, a small distance away from the junction of the header/cantrail/A-pillar). Such possibilities should be considered at an early stage in the PCIV development. Additionally as PCIV vehicles enter service and crash details become available, new or modified testing may be required in response to any perceived weakness in the existing set of regulated tests. Any required modifications to the testing standards will need to be applied to both current conventional steel structures, but more importantly to the evolving metallic designs, which are likely to be as affected as the anticipated PCIV structures.

Specific issues and concerns for PCIVs as related to primary safety standards are addressed in the following sections.

3.2.1 Front and Rear Impact

Although the front and front offset impacts have a far higher public profile than rear impact, the essential requirements from an engineering standpoint are the same; energy absorption and prevention of intrusion.

Both types of testing associated with a front impact, rigid barrier and offset deformable barrier, present a realistic test scenario against which to judge the performance of PCIVs. These tests in general represent actual crash scenarios and as such need no major modification for the PCIV. One possible addition that should be considered is variations in loading direction. For example the offset deformable barrier test could also be conducted at a small angle on incidence (such as 15 degrees). Such testing would serve to highlight any inherent weaknesses in the composite structure associated with the lack of ductility and hence robustness to load case variation.

During a front impact event with a composite energy absorbing structure, energy is dissipated by the destruction of the composite material directly in contact with the impactor. In the laboratory, this event can be produced using a large rigid face attached to the front of a sled. This test setup provides the composite structure with a rigid, flat face against which to crush. In a real crash event, however, this flat face will likely not be present. As the crush of the composite initiates, it is likely that any attachment at the crush front (bumper support attachments, engine mounts, suspension attachment points) will not remain intact. Thus, some form of guided attachments may need to be developed that maintain the attachment of bumpers, crush plates, etc. to the crush structure, but do not inhibit the energy absorbing mechanism.

Composite vehicles are likely to be good at prevention of intrusion into the occupant space, as composite failure modes will not support gross deformation wherever the peak forces are limited by the progressive crush of the energy absorbing structure. Major deformation, or intrusion, into the occupant compartment is a good predictor of injury risk in crashes, even when dummy injury measures are low [72].

3.2.2 Side Impact

The mechanism of side impact energy management relies on energy absorption and anti-intrusion. Although capable of high energy absorption during crushing, composites do not perform well in anti-intrusion once the strength of the material has been exceeded. As this point occurs at low levels of strain, little energy is absorbed, and the anti intrusion mechanism has been destroyed. Alternative design concepts and/or materials will need to be evaluated to address this problem. Some newly created fibers are showing good performance with high elongation to failure. With the proper matrix material (perhaps thermoplastics), such composites may yield the desired ductility while maintaining the required strength, rigidity, and energy absorption required for static and impact loadings. It is noted that the attachment of dissimilar materials may make the task of adhesive bonding more complex.

3.2.3 Roof crush

Roof crush has similar requirements as side impact. Although the FMVSS test is conducted statically, in practice roll-over crashes are highly dynamic events that require a level of energy absorption as well as anti-intrusion. A more realistic test case should be devised that protects the occupant to the same level as a steel roof, but in a more realistic, dynamic manner. Such a test should consider multiple roll-over conditions, including rolling onto a flat surface and one with a load concentrating feature.

3.3 Test and Evaluation Procedures for Composite Materials

Although the detailed relationships between failure modes and the associated energy absorption are not well understood in composites, perhaps this should not be considered a shortcoming in the development of PCIVs. A test procedure may be developed that assesses the suitability of candidate materials for use in energy absorbing structures. Initially, tests should be aimed at identifying materials that display the fundamental requirements of stable crush and sufficient strength for the back-up structure. These tests can be performed using small flat coupons to determine basic parameters, or using more complex element-level specimens such as tubes. Such initial tests may be used to identify candidate material systems for which further testing is required.

In order for the designer to make a reasoned choice of material for any given part of a vehicle and to perform structural analyses, reliable material properties of candidate composite materials must be available. As will be discussed in detail in Chapter 4, characterizing a range of candidate composite materials for use in automotive applications is problematic in comparison to metallics. A significant number of fibers and matrix materials exist from which candidate automotive composites can be composed. Additionally, the properties of the composite can also be affected significantly by the percentages of these constituent materials. Finally, additional specialized crashworthiness properties are needed for composite material to be used in automotive applications where energy absorption is a key consideration.

Chapters 4 and 5 will provide the reader with a detailed description of the progress to date and current status of test procedures for material characterization as well as composite material databases. This section will focus on tests and evaluation procedures that are expected to be required for composite materials to ensure compliance with Federal Motor Vehicle Safety Standards (FMVSS).

3.3.1 Sustained Crush Stress

The sustained crush stress, defined as the average crush load divided by the cross sectional area, is a critical design parameter that must be known when designing an energy absorbing structure using composites. However, the crush stress is dependent on a number of material and geometric parameters. Among the more obvious factors are the fiber type, fiber orientations, resin type, and volume fractions of the constituent materials. Several other factors are less obvious: the overall part thickness, the order and grouping of individual layers, and geometric features such as part curvatures all affect the crush stress and must be well understood.

A simple test method is required to measure the crush stress associated with these variables such that material screening studies may be performed to determine suitable values. As discussed in Chapter 4, test methods are currently under development using small flat coupons to establish the crush stress and energy absorption.

Curvature of a composite component increases the measured crush stress due to the support offered by the hoop tension/compression forces generated in the direction transverse to the crush loading. Typically this support increases the buckling force required to produce crushing, giving a higher crush stress. The effect of curvature can be measured by using a tube test or from sinusoidal-shaped coupons. The effects of curvature can also be investigated using "pin-stabilized" flat coupons which require the specimen to fail in a similar manner to a curved test piece. Such tests have been shown to produce similar values of sustained crush stress to that obtained from sinusoidal-shaped specimens [73].

3.3.2 Compressive Strength

Compressive strength is an important consideration, as it is the primary property used for predicting failure of the composite component behind the crush front. From the compressive strength, the ratio of compressive strength to sustained crush stress, or Compression Crush Ratio (CCR) can be calculated and used as an indicator of the likelihood of the composite material crushing in a stable manner. It would be tempting to suggest that the compressive strength need be some factor greater than the crush strength to provide a desired safety factor. In fact, this safety factor needs to be higher than might be first imagined due to dynamic factors.

Given that a high crush stress or SEA is desirable when selecting materials for composite crush, the designer needs to be aware that too high a crush stress can cause problems with overall performance. For a given cross-sectional area, the higher the crush stress the higher the forces in the crushing component. Almost invariably the limiting factor for successful crushing of the component is its compressive strength. This strength value needs to be high enough to support the crushing forces, including dynamic fluctuations, which cause magnifications to the stress levels seen in the structure.

To illustrate the point, consider material A has a crush stress of 50 MPa (7.3 ksi) and a compressive strength of 250 MPa (36.3 ksi). If a straight rectangular tube of material A does not experience buckling, it has a safety factor of 5 since the compressive strength is 5 times the crush stress. This safety factor is known as the Crush Compressive Ratio (CCR). In reality, the crushing of a component in a vehicle is a highly dynamic event which can magnify the stress condition in the structure, for example doubling the stress seen at certain time intervals. The level of magnification will depend on material damping as well as component geometry. For this reason, the CCR needs to be higher than a pure static analysis of the forces would indicate to be confident of continuous crushing rather than breaking behind the crush front.

For purely prismatic components of suitable wall thickness and side length to prevent buckling, crush can be sustained with relatively low CCRs compared to components with more variable geometry such as may be found in real automotive designs. Real designs are a compromise of many factors, for example a hole in the side of a longitudinal member may introduce a stress concentration factor of up to 3. In such a case, the effective CCR for a material may not be sufficient to prevent failure of the structure.

Composite materials commonly used in motorsport crash members typically have compressive strengths in the region 500 to 800 MPa (70 to 120 ksi), and crush stress levels of 75 to 120 MPa (10 to 17 ksi). These properties give rise to CCRs of 4 to 10 (more commonly 5 to 7). Higher CCRs allow the designer to more easily cope with complex (non-ideal) geometries and load cases such as offset or angled impacts without premature failure in the backup structure. The authors' experience suggests that CCRs below 3 would only be appropriate for ideally shaped components. However, the automotive designer should verify through analysis or testing that the chosen combination of crush stress, compressive strength, material moduli, and geometry will allow the desired crushing behavior to a suitable level of repeatability and reliability.

There are well established, standardized test methods for the measurement of compressive strength. However such tests should be conducted using the same material and manufacturing process as for the sustained crush stress coupons. If using flat coupons for crush testing, it is desirable to obtain both the compression strength specimens and the crush stress specimens from the same test panel.

3.3.3 Interlaminar Shear Strength

Interlaminar Shear Strength (ILSS) is another important property used to predict the strength of the back-up structure as well as to assess the resistance to bending close to the crush front. There are a number of existing tests that measure ILSS in some manner. However, different test methods often produce different test results. While an accurate measure of ILSS is desired for analysis, the consistent use of any ILSS test method may be sufficient for comparing the relative performance of candidate materials. In addition to seeking increased ILSS through material selection, several manufacturing methods have been investigated, including the use of needle punching and through-the-thickness stitching.

3.3.4 Internal Damping

Internal material damping is an important property that is required for use in dynamic analysis. Damping affects the dynamic response of a structure and serves to reduce potentially damaging stress waves travelling through the structure. A measure of material damping is often necessary for crash modeling of composite components. Some numerical codes utilize "sky hook" damping, where the damping forces applied to nodes of an element are a function of their velocity with respect to ground. This approach is sometimes described as performing the simulation in a viscous fluid, or "running the analysis in molasses." While further research is needed to determine the most suitable method of incorporating realistic internal damping, a measure of the internal material damping will remain as an important property for performing crush analyses of composite components.

3.3.5 Strain To Failure

The strain to failure of a composite material may be determined using well established mechanical tests. However, testing to identify composite materials and laminates that achieve high levels of strain to failure may be needed to obtain acceptable levels of anti-intrusion performance. Many of the common fibers and resins used in structural composites have low strain to failure values and behave in a relatively brittle manner upon impact. For best anti-intrusion performance, the ability to retain structural performance after high strain or initial/partial failure will give designers the best materials to achieve their goals of vehicle safety. Many of the higher strain-to-failure fibers have been found to exhibit relatively poor interfacial bonding, reducing their capability in tension and shear.

3.3.6 Mechanical Damage

Mechanical damage may be inflicted during normal use of a vehicle. This damage, whether produced from low-speed impacts, from road debris kick-up, or other sources, can take the form of fiber breakage, matrix damage, and/or interlaminar damage. Such damage, if of critical severity and/or in a critical location, can lead to premature failure of the back-up structure during a crush event. If present at the crush front, such damage may be tolerable, reducing the energy absorption by only a small amount.

The degree of damage sustained for any given event can be difficult to determine visually and some form of non-destructive inspection may be required. However, analyses may be used to identify regions of a structure where damage may be critical as well as areas that are prone to such damage. Protective coatings or covers offer one possible solution to the formation of such damage.

An important consideration in the design and development of a future PCIV concerns the ability to repair composite structures to an acceptable level for subsequent redeployment. Composite monocoques structures are able to be repaired in Formula 1 racecars [34]. The likelihood is that PCIV structures will be less complex in their construction and due to the multiple loading and durability demands likely to have thicker sections which will be easier to repair. Nevertheless, any new design will need to be able to demonstrate which areas of the structure are suitable for repair and which areas is repaired will change the crush response of the vehicle. Further, vehicle developers will need to consider whether a repair to an energy absorbing structure will crush progressively without increasing the loads entering the safety cell to a level where degradation and instability may be initiated.

3.3.7 Environmental Effects

Composites are in general very tolerant to environmental effects such as UV damage, moisture, chemical attack, and temperature extremes. In the material selection process, however, it is important to consider that if a composite material experiences sufficient degradation due to one or more of these effects, appropriate measures should be taken or the material should not be considered further. Additionally, the effects of the environment in service operating conditions should not adversely affect the performance of the vehicle in an impact later in the life of the vehicle. It is worth keeping this in perspective with the challenges related to the corrosion of steel vehicle structures.

3.3.8 Adhesives

Joints represent one of the greatest challenges in the design of lightweight composites structures. There is a significant cultural barrier to introducing adhesively bonded structures to high volume automotive components because of the lack of history and experience in this area. Welding components together is inexpensive and has been developed by the automotive industry over many decades. In principle, adhesive joints are structurally more efficient than mechanically fastened joints because they provide better opportunities for eliminating stress concentrations. However, adhesive joints tend to lack structural redundancy, and are highly sensitive to manufacturing deficiencies, including poor bonding technique, poor fit of mating parts and sensitivity of the adhesive to temperature and environmental effects such as moisture. In order to make them effective, detailed design and analysis is required to prevent local effects initiating failure in the relatively low loaded bulk adhesive of the joint.

Adhesives will play a crucial role in the PCIV, as adhesive bonding is an ideal way of joining composite panels. Good load spreading and sealing are two advantages of using adhesives to join composite components to other parts of the vehicle. There are, however, some obstacles for the use of adhesives. In line with many other resin-type materials, the relative lack of ductility would be an area that would need careful consideration for any specific application. The lack of ductility can be problematic for both the fatigue resistance and for safety in an impact. Some regular epoxy-type adhesives are prone to crack initiation and propagation can lead to degradation of the structure. Additionally, rapid crack growth can occur during impact (initiated in the crushing zone) leading to rapid failure of the bonded joint and hence failure of the entire energy absorbing structure. One possible solution is through the use of mechanical fasteners to inhibit the crack growth at the joint and retain structural integrity in the case of adhesive failure. When positioned appropriately these tend to prohibit crack initiation, can arrest cracks, and prevent rapid crack growth. This is achieved by eliminating the tensile component across the adhesive joint at the point of application. It should be noted however that mechanical fasteners that require holes drilled or pierced through the composite give rise to potential fatigue initiation sites, requiring great care in their positioning. However, fasteners that do not require holes but that are crimped across an adhesive flange would eliminate this limitation if such a method proved to be sufficiently strong and stiff to resist the peel forces opening the crack. Perhaps the largest resistance to the use of mechanical fixings would be the added cost to an assembly. A more cost-effective solution would be an adhesive that does not suffer these limitations, and indeed some currently available adhesives offer considerably higher strain-to-failure values and fracture toughness.

Mechanical properties of adhesives are currently obtained using a variety of shear and peel tests [29]. However, many of these properties are dependent on the geometry of the test specimen and as such are not true material properties. The use of such properties for adhesives in finite element analyses of adhesively bonded components is questionable, since the geometries do not allow direct comparison. Suitable tests need to be performed such that intrinsic material properties and failure criterion can be deduced for the adhesive to allow the analysis of bonded joints to be conducted reliably [74].

Another consideration regarding the use of adhesives is the temperature range to which the bonded structure will be subjected. Typically, temperature effects are not considered for steel vehicles, and testing is typically performed at ambient conditions. For many epoxy-type adhesives, however, the strength, stiffness, and ductility are affected by temperature. The possible effects of temperature on the properties of adhesives should be considered when assessing the crash performance of adhesively bonded structures. Furthermore, joint integrity is often compromised for joints containing dissimilar materials, as the difference in Coefficient of Thermal Expansion (CTE) must be accommodated through strain in the joint at service temperatures that differ from the cure temperature.

3.3.9 Joint geometry

Considerable efforts are being made in the chemistry of adhesives in order to improve the strength, durability and impact resistance of adhesives. However it should not be overlooked that an important part of maintaining the joint integrity of a bonded assembly is the geometry of the joint itself.

Flat flange joints used for joining sections to create a box section, for example, are prone to rapid crack growth through the adhesive joint under impact conditions. In a box section with joints as shown in Figure 3-1a, cracks in the flange joints can grow to the full length of the section within the first 50 mm (2 in.) of crush. This of course leaves the section very vulnerable to instability and poor energy absorption. In addition to solutions discussed above, there are a number of possible designs that also contribute to the robustness of the joint. As shown in Figure 3-1b and Figure 3-1c, the use of multiple faces set at an angle (nominally 90 degrees) to each other can create peel resistant joints. In Figure 3-1b as peel forces create crack growth in the vertical part of the flange, the horizontal part limits the degree to which the joint can open and hence limits the length of the crack to several centimeters in front of the impact face. Thus the sections stability is maintained during the impact event. The section in Figure 3-1c behaves similarly, which would be dependent on the sequence of events in a given impact scenario.

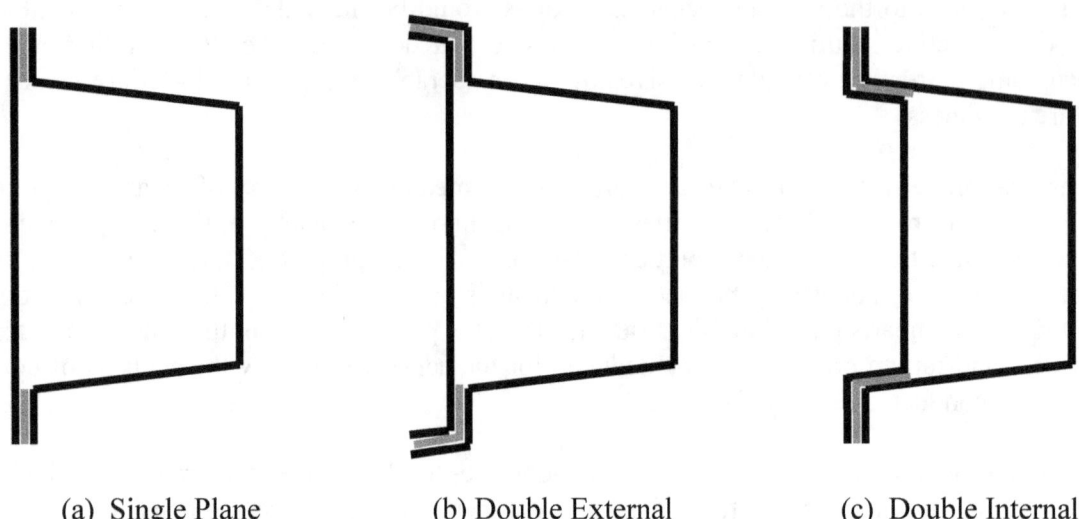

(a) Single Plane   (b) Double External   (c) Double Internal

Figure 3-1. Conventional and Peel Resistant Joint Geometry.

3.3.10 Crack Arresters at Bonded Joints

Crack growth through a continuous bead of adhesive is a mechanism that can lead to catastrophic failure of the joint in fatigue but especially in impact situations. However, such crack growth could be arrested by the use of non-continuous beads of adhesive. In such a joint, as a crack develops and propagates along the length of an adhesive bead, it would stop at the boundary of the adhesive bead and not propagate to the adjacent bead. For the crack to continue, a new crack front would need to be created at the adjacent bead of adhesive. This sequence of crack initiation, propagation, and arrest leads to a higher resistance to run-away crack growth [75]. Determining the size and spacing of such adhesive beads to resist crack growth will require future research, and the types of loading and properties of the adhesive must be taken into consideration.

Enhancing the fracture toughness of adhesive bonds is a desire across many industries. Research into the joining of wood substrates using droplet dispersion for the purpose of increasing fracture toughness, is of direct relevance for the improvement of composite joints in future PCIV structures [76].

The aircraft industry has considered the need to enhance the fracture toughness in repair patches for aircraft structures in order to arrest the development of fatigue cracks and where the loads inevitable lead to a failure that the patches locally debond and in the process give an indication of structural damage during routine maintenance [77].

3.4 Evaluation Procedures for Composite Designs and Components

There is a fundamental difference in the way composite materials absorb energy in a crash versus metallic structures. Energy absorption in a metallic structure is characterized by plasticity and folding of the material without gross material failure. In contrast, composites absorb energy through material fragmentation, leading to the destruction of the part. As a result, the evaluation procedures used in the design of composite automotive components differ from those used for metallic structures. Aspects of evaluation procedures which differ for composite automotive components are detailed in this section.

3.4.1 Failure Modes

A general understanding of the failure modes produced in composite materials under consideration is important for designing vehicle components with the desired energy absorption and/or anti-intrusion characteristics. Some parts or subsystems may require both characteristics. For instance, the side panels of an automobile require good anti-intrusion capability to protect the occupants from an impacting structure, but also a measure of energy absorption to allow for energy management during the deceleration phase of the impact event.

The energy absorbing mechanisms present in composite materials during crush is currently not well understood. Energy absorption results from many possible mechanisms in varying degrees depending on the particular material system. An understanding of these mechanisms and how to "activate" them to achieve the desired energy absorption characteristics is needed. Additionally, testing is needed to assess several key characteristics that allow comparison between materials with a view toward material selection and structural analyses.

3.4.2 Design Evaluation

Conventional crash structures are based around linear crash members for front and rear crash energy absorption. Recent efforts in researching the crush performance of plastics and composites have largely been based on developing the composite equivalent of a metallic linear energy absorbing member. Even though the energy absorbing mechanism is different (i.e. fragmentation rather than plastic folding and buckling), the macroscopic goal is that a linear member, aligned with the direction of vehicle velocity, will provide a resistive force to slow the vehicle in a progressive manner. One shortcoming of the linear crash member design (whether in metal or composite) is that it is optimized for one loading direction and often performs comparatively poorly when subjected to loads in other directions. As an incremental development, it may be possible to design composite versions of the linear crash member that exhibit high energy absorption under multiple loading directions. For example rather than two longitudinal members it may be possible to design a composite structure with a number of load paths, offering structural redundancy (i.e. if one load path fails then others can carry the additional load) and progressive crushing under a range of loading directions.

While future PCIV structures are currently not well defined, the number of individual panels making up a Body in White (BIW) will be considerably lower than that of a contemporary steel BIW structure. As an example, the Automotive Composite Consortium's Focal Project III Composite BIW consisted of only sixteen panels as opposed to the 240+ panels in the baseline Chrysler Cirrus [78]. In addition, the steel panels are connected by spot welds (and continuous welds where reinforcement is necessary) with little if any bonding of materials. On the contrary, the PCIVs of the future will have a significant reliance on adhesive bonding of panels. Many of such bonds will be multi-substrate and of variable thickness to allow for panel tolerance. The adhesives being deployed in recent composite BIW structures span a range of ductility and stiffness requirements, depending on the application. In addition, the bond thicknesses required range from contact through several millimeters. Future PCIVs will require an understanding of the performance of the adhesives under various loading regimes in new joint configurations, and with multiple substrates.

3.5 Evaluation Tools for Composite Designs and Components

In line with the development of steel vehicles, the main tool for the development and evaluation of composite vehicles will be the use of Finite Element Analysis (FEA). This method allows for many design iterations to be tested virtually without the need for constructing physical parts. Due to the differences in energy absorption modes between composite and steel structures, the existing, highly successful non-linear finite element codes used for the development of metallic vehicles are unable to predict the crush of composites in their current form. As will be discussed in Chapter 4, however, there are a number of methods being developed to predict the crush response of composites. In general, these FEA methods for composite structures remain under development.

The inputs to these finite element codes consist of structural geometry, material properties, loading conditions, and constraints. For the analysis of composite components, sensitivity studies focusing on both material properties and loading conditions must be considered to ensure a robust design. These factors are discussed in the following sections.

3.5.1 Material Properties

Material properties should be drawn from a reputable database or obtained using recommended test methods. Additionally such data will include a measure of the degree of scatter, which may be used for sensitivity analysis. In such analyses, a range of possible properties should attempt to cover the worst combination of properties to ensure that the component is robust under all such likely occurrences. For example, the analysis of a front longitudinal energy-absorbing member would require the use of the lowest energy absorbing crush value to ensure that there is sufficient crush length to absorb the required energy. Additionally, an analysis performed using the highest crush stress in combination with the lowest static strengths in the back up structure would be considered to ensure that there will not be a premature failure in the back-up structure. A future development of this process might be stochastic analysis, where the statistical variation found during testing would be used to introduce random variation in the properties used in the analysis. This approach is believed to be capable of delivering a robust design, capable of fulfilling the design goals while addressing all likely tolerance variations.

3.5.2 Loading Conditions

Similarly to the variation in possible in material properties, there exists a variation in possible loading scenarios. This variation is believed to be more important to consider for composites, since energy absorption of steel structures through yielding and plastic deformations is inherently more tolerant of loading variations than energy absorption through brittle fracture of composites. The degree to which the loads may need to be varied may initially be prescribed by the designer, but in time should be the result of investigation of further research into crash statistics and prescribed by those certifying automotive safety.

3.6 Test Procedures for Composite Designs and Components

As part of the design process, there are times when it is appropriate and highly desirable to test specific components or subsystems in isolation from the complete structure to ensure that the behavior under test is as expected. This process allows smaller incremental steps forward and is especially important for breakthrough designs that use new materials, such as first-generation PCIV vehicles.

When testing composite components as part of the design process, it is important that the test articles be manufactured to be as representative as possible to the production parts, both in materials and manufacturing process. This includes the fiber and resin architecture, material application, and the curing cycle. These factors can have a pronounced effect on the properties of the component and hence on its performance in test.

When testing a component in isolation from the rest of the design, it must be attached to the support frame in such a manner as to not influence the test outcome. When testing an energy absorbing structure, care must be taken to not over-constrain the structure and lend additional support to buckling-prone components. Such over-constraint may lead to a positive test result that conceals a fundamental design flaw. Similarly, under-constraining the test article could have the opposite effect, resulting in a good design being eliminated from consideration.

Equally important as the support frame is the manner of loading. Stability of the applied load as well as the load application rate may influence the final test result. Monitoring the test during loading is crucial to gain a full understanding of the mechanisms at work, especially for dynamic tests. High-speed video taken from multiple angles during the test is highly desirable. Additionally, care should be taken to collect test data at an appropriate frequency as to not miss significant test article responses.

Following crush testing, a post impact reconstruction of the test article can be very informative. Since composite structures absorb impact energy through fragmentation, a significant amount of dust and debris typically is created during the impact event. This dust and debris can obscure the view of the high-speed video cameras, making an accurate determination of the crush sequence difficult. In such cases, the post-impact event reconstruction of the test article can assist in assessing the failure process. One method adopted by Barnes [79] to assist in test article reconstruction involves marking and/or painting the article in a descriptive manner. On more complex structures, individual panels can be painted different colors and some location reference markers provided on the outer surfaces to an appropriate resolution.

A potential issue with the testing of composite components, especially when manufactured with material forms that demonstrate relatively high levels of variability in material properties, is the repeatability of test results. For such materials, testing should be performed on a suitable number of replicates to ensure that the variability in the performance of the component is well understood. Similarly, testing to determine the sensitivity of a component to loading conditions should be conducted. Such testing would include variations from the nominal test conditions, including the angle of load application and the loading rate.

3.7 Conclusions and Recommendations

The current FMVSS tests have been developed over many years as a result of practical experience of real-life crashes. As such, these tests should be used as the fundamental basis for PCIVs and other light weight vehicles. In the development of future FMVSS tests and component tests, it is recommended that such tests adequately assess the ability of a design to protect against injury rather than to meet the requirements of a controlled environment. Additionally it is recommended that any future FMVSS test standards designed to address PCIVs are also applicable to current metallic vehicles and any other future lightweight alternatives.

4. SUMMARY OF PROGRESS IN MATERIAL DATABASES, TEST METHOD DEVELOPMENT, AND CRASH MODELING

4.1 Introduction

In this chapter, progress is summarized in three topic areas pertinent to crashworthiness: material databases, crashworthiness test method development, and crash modeling. This chapter focuses on the evolution of research and development efforts in each of these three areas, leading to current research activities. Current research activities and a summary of the current status in each of these three topic areas are presented in the following chapter.

4.2 Summary of Progress: Material Databases

Obtaining standard stiffness and strength, or "mechanical" properties of composite materials, has traditionally been a problem for designers/analysts of composite structures. The nature of fiber-reinforced composites – consisting of a specific type of reinforcing fiber and a matrix material – makes for a large number of resulting composite materials. Additionally, the percentage of the fiber and the matrix materials within the composite can vary, depending on the processing methods used. These percentages of the constituent materials in the composite, often reported as *volume fractions*, can vary significantly when processed using different methods. Different composite materials with significantly different mechanical properties can be produced from the identical fiber and matrix materials. While some mechanical properties can be extrapolated to different volume fractions of fiber and matrix through micromechanics analyses, others require mechanical testing to establish their sensitivity to changing volume fractions of fiber and matrix.

An added complication to establishing standard stiffness and strength properties for composite materials arises due to the fact that these properties are direction-dependent. Whereas a metal or plastic is often assumed to exhibit isotropic material behavior (properties are the same in any direction), the direction-dependent nature of fiber-reinforced composite materials requires that testing be performed to obtain properties in two different in-plane directions of a unidirectional composite – or composite "lamina." Additionally, strength properties of a composite material cannot be assumed to be the same under tension and compression loading due to different failure modes produced and separate tests must be performed. Thus, the number of tests required to obtain the most basic stiffness and strength properties of a composite material is relatively large and expensive to perform. Once these material properties are obtained, they are only applicable to a specific composite material – consisting of a specific type or "grade" of reinforcing fiber, a specific matrix material (such as an epoxy), specific processing conditions, and a specific volume fraction of each constituent. As a result, it is not uncommon for a designer/analyst to be required to consider composite materials for which a complete material database is not available.

In contrast to composites, obtaining the stiffness and strength properties of plastics and metals is simplified by their isotropic nature, possessing the same properties in all directions. Additionally, it is often assumed that the stiffness and strength properties of these materials are the same in tension and compression, further reducing the number of required tests. Although there are numerous metallic alloys and plastics that may be considered for a design, the more conventional choices tend to be relatively standard materials, with well established material databases. While it is relatively uncommon to find well-populated material databases for composite materials, they are commonly available for metals and plastics. Additionally, the properties required to simulate crash performance of metallic automotive structures are relatively basic, generally being limited to the stiffness properties, yield stress, elongation behavior, and rate dependency.

An added complication for locating material databases for use in automotive composite design is that the relatively small number of composite materials with commercially-available, well-populated material databases have been developed for use in the aerospace industry. Such composite material systems typically are composed of high performance and relatively expensive fibers and resins, manufactured using an aerospace-type manufacturing method (such as autoclave curing) that is capable of producing high fiber volume fractions and higher stiffness and strength properties, but at a higher cost. Such manufacturing methods, while well suited for relatively low-volume, high-performance applications, are generally not well suited for cost-conscience, high-volume automotive applications.

The opportunity within the automotive industry for composite material suppliers and manufacturers is vast. A PCIV vehicle produced in volumes of 100,000 vehicles would account for approximately 50% of the projected demand for carbon fiber in 2014, or approximately 64 million kilograms [80]. With these indicated volumes the overhead to produce detailed material databases on candidate composite materials would be relatively low.

4.2.1 Sources of Material Databases for Composites

Perhaps the most obvious source of material data for any engineering material is the material supplier. For composite materials, this "supplier" is often thought of as the producer of the final fiber/matrix composite. Typically producers of composite materials provide limited, "representative" material properties for their composite materials. The most common material properties provided include the 0° and 90° modulus and strength under tensile and compressive loading; in-plane shear modulus and shear strength; flexural stiffness and strength; and short-beam shear strength. Somewhat less common properties provided by material suppliers include the open-hole compression strength and compression-after-impact strength. Typically the material properties given are average values without statistical information. Additionally, values provided may be in the upper ranges of what is achievable or from ideal processing conditions. Since mechanical properties vary with fiber volume fraction and become dependent on the fabrication process, typical material supplier data may be of use for initial candidate material selection. However such data is typically not adequate for the design and structural analysis of composite structures or components.

Although more extensive databases have been developed for several composite materials, they are generally not publicly available. A majority of these databases are either company proprietary or have restricted distributions due to government regulations. These databases often are generated for a specific composite program or structure, and the extensive data generated provides design allowables for a specific composite material manufactured using a specific process. Further complication arises due to large aerospace corporations using their specific test methods. In general, these company-specific or program-specific composite material databases are not publicly available.

The first major effort to produce an extensive shared database for multiple composite materials began in late 1994 with the creation of the Advanced General Aviation Transport Experiments (AGATE) program, sponsored by NASA and the Federal Aviation Administration (FAA). One of the primary focuses of the AGATE program was to create shared data bases for specific composite materials of interest for composite aircraft components in the general aviation industry. With the AGATE database, a developer of composite aircraft components could select from the list of characterized composite materials, and significantly reduce the time and cost associated with material characterization. In total approximately 70 member organizations from industry, government agencies, and academia participated in AGATE, and material databases were generated for several composite materials of interest to the general aviation community. The composite materials characterized included both carbon fiber and glass fiber reinforced composites composed of both unidirectional fiber orientations and woven fiber fabrics. Composite materials characterized included both those manufactured using aerospace-type autoclave curing processes as well as lower cost vacuum-only curing processes. Although the AGATE program officially ended in 2001, some material suppliers continued to add new composite materials to the shared database afterwards.

In 2005, the National Center for Advanced Materials Performance (NCAMP) was formed at the National Institute for Aviation Research at Wichita State University. One of the primary functions of NCAMP was to extend the material characterization efforts of the previous AGATE program to the entire aerospace industry [81]. Through NCAMP, material databases continue to be developed for new composite material systems. In general, these additional composite materials may be classified as high-performance composite materials, intended for use in the aerospace industry. As a result, the composite materials that have been characterized under AGATE and NCAMP utilize high-performance and therefore high cost materials and processes, and are of limited usage for high production usage in the automotive industry.

4.2.2 Specialized Crashworthiness Properties

In addition to the mechanical properties discussed in the previous section, additional properties are needed for composite material to be used in automotive applications where energy absorption is a key consideration. Two primary properties are of interesting for assessing crashworthiness of a composite material in a particular application: the specific energy absorption, and the sustained crush stress. Each is described briefly below.

The first specialized crashworthiness property desired is a measure of the energy absorbed during crushing. Typically, a crushing event begins with a rapid rise in force until the maximum, or peak, compressive load is achieved. Thereafter, the desirable type of post-peak performance, known as *progressive crushing*, is characterized by a relatively constant load that is typically less than the peak load. The resulting force versus displacement plot characteristic of progressive crushing is shown in Figure 4-1. From this load versus displacement response, the Specific Energy Absorption (SEA) may be determined. The SEA is defined as the energy absorbed per unit mass of crushed material, and can be written as

$$\text{SEA} = \frac{W}{\rho A \delta} = \frac{\int_0^\delta F d\delta}{\rho A \delta}, \qquad (1)$$

where the total energy absorbed, W, is equal to the integral of the load, F, over the total crush displacement δ, or the area underneath the load versus displacement curve. The quantity ρ is the material density, and A is the cross-sectional area of the specimen. SEA is widely believed to be dependent on the strain rate in the composite material during crushing, and thus SEA results obtained from quasi-static compression testing may not be the same as those obtained from dynamic crush experiments that produce higher strain rates in the material.

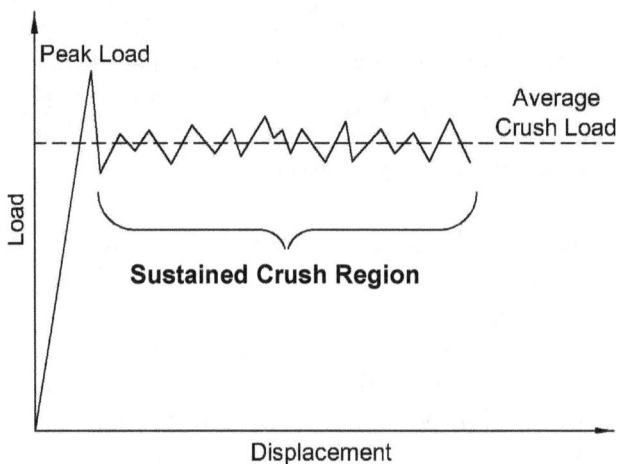

Figure 4-1. Typical load versus displacement plot obtained from progressive crushing of a composite test specimen.

A second quantity of primary interest is the sustained crush stress, defined as the average crush load (as shown in Figure 4-1) divided by the cross sectional area, A, of the specimen. The sustained crush stress is of particular interest when compared to compression strength (the initial peak compression load divided by the cross sectional area, A) for establishing the percentage of the compression strength of the test article at which progressive crushing will occur. Similar to the SEA, the sustained crush stress is believed to be dependent on the strain rate in the composite material during crushing.

From the sustained crush stress, the Compression Crush Ratio (CCR) may be obtained. The CCR is defined as the ratio of the compression strength to the sustained crush stress of a composite laminate. While serving as an important metric to indicate the likelihood of stable crushing, the CCR does not constitute an additional crashworthiness property.

Additional properties of interest in analyzing crashworthiness include the fracture toughness and material damping associated with a composite test article. Both properties are of interest for finite-element based crashworthiness modeling of composite structures. Other characteristics of interest are the failure modes and damage progressions associated with the crush event. These characteristics are of interest for establishing an understanding of the sources of energy absorption in composites as well as for the validation of finite element model predictions.

Of the specialized crashworthiness properties described above, only the fracture toughness has appeared in any of the material databases described previously for mechanical properties. However, the Automotive Composites Consortium (ACC) has a material database under development for specialized crashworthiness properties of composites. This database, currently in its final development stages, contains data generated by ACC research activities, will be available to ACC participants but not to the general public. Otherwise, the source of such specialized crashworthiness test results has been limited to journal articles, research reports, and conference proceedings related to crashworthiness testing. Such publications have, in general, focused on the development of crashworthiness test methods and are not considered as material databases for specialized crashworthiness properties. A review of crashworthiness testing efforts is provided in the following section.

4.2.3 Summary: Material Databases

Table 4-1 summarizes the current material databases for composite materials. Included in the summary table are the availability of the databases and their relevance to crashworthiness of composites. Recommendations concerning the development of material databases for composite crashworthiness are presented in Chapter 5.

Table 4-1. Summary of Material Databases For Composite Materials.

Database Provider	Data	Usage	Relevance to PCIVs	Availability
Material suppliers	Limited stiffness and strength properties	Initial selection of candidate materials	No indication of crashworthiness	Available to public
Company-specific or program-specific database	Extensive stiffness and strength properties	Design of aerospace structures	No indication of crashworthiness Aerospace processing and materials	Company proprietary, limited distribution
AGATE and NCAMP characterization programs	Extensive stiffness and strength properties	Design of aircraft/ aerospace structures	No indication of crashworthiness Aerospace materials and processing	Limited availability
Automotive Composites Consortium (ACC)	Data generated from ACC research activities	Initial crashworthiness assessment Model validation	Focus on crashworthiness properties Test methods not standardized	Available only to ACC participants

4.3 Summary of Progress: Crashworthiness Test Method Development

Crush testing is the primary means by which the Specific Energy Absorption (SEA) and crush behavior of composites is evaluated. Testing is also foundational to building empirical relationships and computational models for use in the design of crashworthy composite automotive structures. When testing composites for crashworthiness, two general test methodologies may be followed. The first methodology involves the use of a test article that is intended to be "representative" of the intended application. This methodology is often referred to as "element-level" or "structural" testing. In general, such test methodologies utilize self-supporting, structure-like geometries. In contrast the second methodology, referred to as "coupon-level" testing, uses a relatively small test coupon that does not contain structural-level features that may be found in the intended application. As discussed in Chapter 1, both coupon-level and element-level testing are the focus of the first level, or base, of the Building Block approach, which focuses on assessing material behavior. Whereas coupon-level testing may be required for material/laminate screening as well as to obtain crashworthiness-specific properties for computational analyses, element-level testing is used when the geometry of the test article is intended to be "representative" of the intended application.

Several test methodologies have been investigated at both the coupon-level and element-level in an effort to characterize the energy absorption capabilities of composite materials and structures. Currently, however, there is no standard by which either type of test method may be performed. In this section, a summary of progress in the development of both types of test methods is presented. First, however, two aspects of crashworthiness testing common to both coupon-level and element-level test methodologies are discussed: classifications of composite crushing, and the use of crush initiating triggers.

4.3.1 Classifications of Composite Crushing

The crushing phenomena in composites can be classified into three types, as has been summarized by Xiao [82] with reference to tubes and by Barnes [83] with reference to flat plates.

Type 1: This failure mechanism is characterized by fiber and matrix fragmentation, resulting in relatively small crush debris as shown in Figure 4-2a. In engineered fabrics or unidirectional reinforced sections, this mode yields high SEA results. This mode would be clearly accepted as crush by the consumption/disintegration of the material occurring at the interface with the impactor. The force levels are usually more consistent with this crush mechanism.

Type 2: This failure mechanism is characterized in flat coupons by significant delamination ahead of the impactor as shown in Figure 4-2b. Such delaminations tend to increase in length as the crush speed increases from quasi-static to above 1.0 m/s (3.3 ft/s) for most composite materials. This mode leaves the fibers largely intact but the resin significantly fragmented. In the tube impact case, this failure mode would be characterized by the formation of "fronds". These fronds are directly analogous with the flat coupons, and have virtually no residual structural capability. In the tube case where the corner tearing occurs between the fronds, additional failure mechanisms are at work.

Type 3: In this classification (Figure 4-2c), the failure mode is essentially not crush. For some composite tubes (such as the Kevlar fiber composite tubes in [82]), the failure mode is folding and concertinaing in a similar manner to that observed in ductile metallic structures. This type of failure is believed to result from the high tensile strain-to-failure nature of Kevlar-reinforced composites. A key characteristic of this type of failure, observed in both tubes and flat coupons, is that significant energy absorption occurs due to bending failure occurring away from the crush front. For composites with lower strain-to-failure (glass and carbon fiber composites), a similar failure also occurs away from the crush front, but in a more brittle manner, with essentially undamaged chips of material produced between the "fold" lines [83].

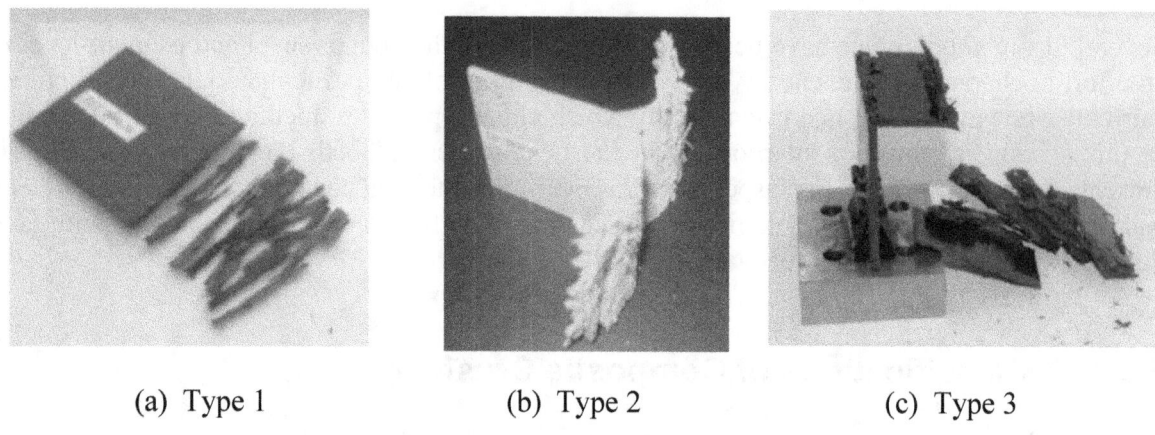

(a) Type 1   (b) Type 2   (c) Type 3

Figure 4-2. Classifications of composite crushing.

4.3.2 Development of Crush Initiating Triggers

Regardless of the crush test methodology followed, composite test articles often require a crush initiating trigger. Triggers promote progressive crushing and preempt catastrophic failure by providing a localized stress concentration, which removes the initial force peak observed if no initiator is present. One form of crush trigger used for composite specimens is a geometric feature located on one end of the specimen. Several trigger geometries that have been investigated are shown in Figure 4-3. Researchers have determined that both the size and the geometry of the trigger can influence the resulting energy absorption during crashworthiness testing [84-86]. The bevel trigger, shown in Figure 4-3a, is a width-wise chamfer machined across one end of the specimen. The steeple trigger (Figure 4-3b) consists of two adjacent bevels with a common apex is at the center of the specimen thickness. Both the bevel and steeple triggers with their apex parallel to the faces of the specimen. In contrast, the notch or "serrated" (Figure 4-3c) and tulip (Figure 4-3d) triggers have apexes perpendicular to the faces of the specimen. Since delaminations between plies of composite laminates naturally orient themselves parallel to the faces of the specimen, the use of the bevel and steeple triggers (Figures 4-3a and 4-3b, respectively) have been found to be more prone to producing delaminations and result in lower energy absorption than the notch and tulip triggers (Figures 4-3c and 4-3d, respectively) [84, 85, 87]. When using the notch trigger (Figure 4-3c), however, crushed material is discharged from both sides of the coupon center line. The failure mode is characterized by fragmentation of the impacting surface by multiple failure modes including matrix cracking, fiber microbuckling, and ply delamination. Typically, the crushed material is totally destroyed, the debris exhibiting considerable amounts of dust and small particles. Thus, the crushed material has little inherent strength remaining due to the comprehensive state of damage. Of the four trigger configurations shown in Figure 4-3, the bevel trigger has been used most frequently [86, 88-90], perhaps at least partly due to its simplicity. This trigger configuration is generally viewed less favorably for use with flat-coupon test specimens due to the production of delaminations as discussed. For flat specimens, the notch trigger appears to be a relatively common choice, especially in recent investigations [73, 91].

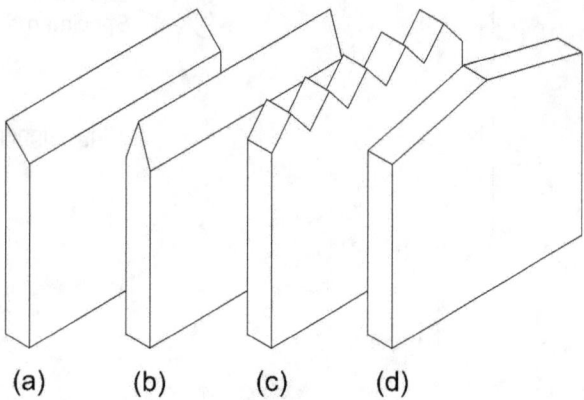

Figure 4-3. Crush triggers used in composite specimens: (a) bevel, (b) steeple, (c) notch, (d) tulip.

In addition to the specimen-based crush triggers shown in Figure 4-3, a crush trigger may be designed into the test fixturing as illustrated by the plug trigger over which a tubular coupon is crushed in Figure 4-4. As the tube is driven onto the radius of the trigger, the tube is stretched circumferentially, leading to axial tearing. For square tubes, such tearing occurs at the corners of the tube whereas for round tubes, tearing occurs at various sites around the circumference of the tube. As the material tears, strips of material between the tears, or "fronds" are created. These fronds are forced around the radius of the plug, causing bending failure in the composite material. The fronds often have residual strength and lower SEA values are achieved, due to the lack of comprehensive crushing of the material. The plug crush trigger does not initiate a Type 1 or Type 2 crush behavior (as classified above), but rather a continuous bending failure as the fronds are driven onto the radius of the plug. The degree of bending failure and the associated SEA is dependent on the radius used on the plug trigger. As a result, the notch-type trigger (Figure 4-3c) and the plug trigger (Figure 4-4) can produce different failure mechanisms. Warrior [92] was able to produce a higher value of SEA using a plug trigger, but only using a specific ratio of plug radius to tube thickness ratio and also high strain to failure matrix and fibers. Otherwise, most literature points to a lower SEA with the plug trigger. In general, however, tube testing performed using plug triggers produces lower values of SEA than flat-coupon tests using notch initiators due to the lack of Type 1 crush behavior. Additionally, plug triggers cannot be used with tapered or variable thickness tubes.

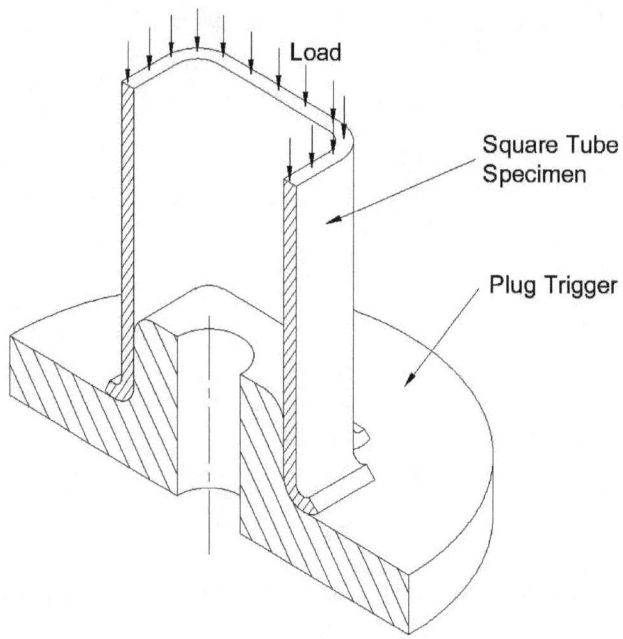

Figure 4-4. Section view of a standard plug trigger used with a square tube specimen.

One noteworthy variation of the standard plug trigger was that of Brimhall [93] in an attempt to isolate the energy associated with the corner tearing that often occurs during the crushing of square tubes. For these experiments, a long, tapered plug trigger was used, which acted to split the corners of the tube without inducing bending and crushing of the flat sections of the square tube. Although most often utilized with tube specimens, fixture-based crush triggers have also been used by Stapleton and Adams [94] when performing edgewise crush testing of composite sandwich panels.

Another method of initiating crush in a composite tube is through the use of a closed-end feature. Barnes [83] has investigated the use of a closed end on a rectangular tube with a thickness of approximately 2.5 mm (0.10 in.) and a 10 mm (0.40 in.) radius to the closed end as shown in Figure 4-5. This closed end also may be used to facilitate the attaching of the tube to another structure, such as an automobile bumper.

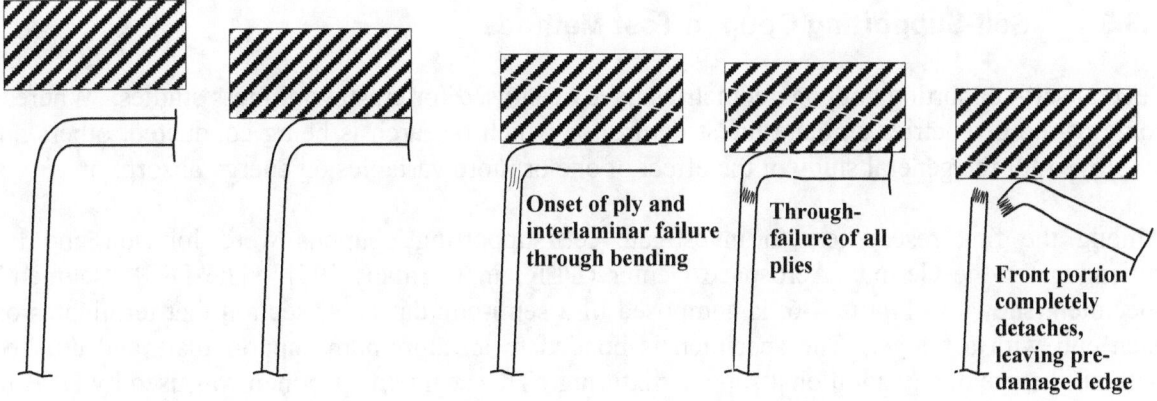

Figure 4-5. Initiating crush using closed-end feature [83].

4.3.3 Coupon-Level Test Methods

Coupon-level test methodologies represent the starting point in the Building Block approach for developing composite structures for crashworthiness. Such coupon-level test methods may be divided into two categories, based on the general shape of the coupon. Test methods for "self-supporting" coupons incorporate a coupon with out-of-plane curvature. If the coupon dimensions are chosen judiciously, self-supporting coupons do not require external supports to prevent global buckling and catastrophic failure. In contrast, test methods that utilize flat coupons require specialized fixtures to achieve stable crushing. Flat coupons are appealing for crush testing of composites because they can be fabricated quickly and inexpensively when compared to other shaped coupons. While flat coupons are not representative of the geometry found in common structural components, such structures commonly have regions of flat geometry. Thus, results from flat-coupon crush testing may be applicable to such regions of a structure. Additionally, flat coupons are useful for studying laminate characteristics, making relative comparisons of composite materials and fiber architectures, obtaining input data for computational models, and may potentially be used to predict aspects of structural behavior.

4.3.3.1 Self-Supporting Coupon Test Methods

Several self-supporting coupon geometries have been used for crashworthiness studies. Whereas some of the geometries emulate the structure for which research is being conducted, others are employed for the general study of the effect of one or more variables on energy absorption.

Among the first researchers to investigate self-supporting coupons were Johnson and his colleagues at the German Aerospace Center (DLR) in Germany [95]. The DLR "segment" specimen, shown in Figure 4-6, is composed of a semi-circular cross section that terminates on each end with a flange. The specimen is bonded to an aluminum support plate and crushed without the use of any additional support fixturing. This segment specimen was used by DLR to compare the SEA of selected material systems under both quasi-static and dynamic load rates.

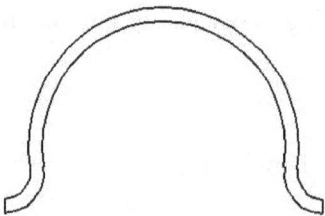

Figure 4-6. Cross section of DLR segment coupon [95].

The most commonly investigated self-supporting coupon geometry has been the sinusoid-like shaped specimen, referred to as the "sine wave" specimen or the "corrugated" specimen [86, 90, 96] and shown schematically in Figure 4-7. The cross-section of the specimen is a series of circular arcs. Geometric variables involved in the sinusoidal specimen include the coupon width (total number of waves), the gross thickness, the width-to-gross thickness ratio, and the included angle [86]. One attraction of the sine-wave specimen has been the similarity of the specimen geometry to actual structural elements used in airframe structures [96]. The cross section of the corrugated specimen used in the CMH-17 round robin investigation [90] is shown in Figure 4-8.

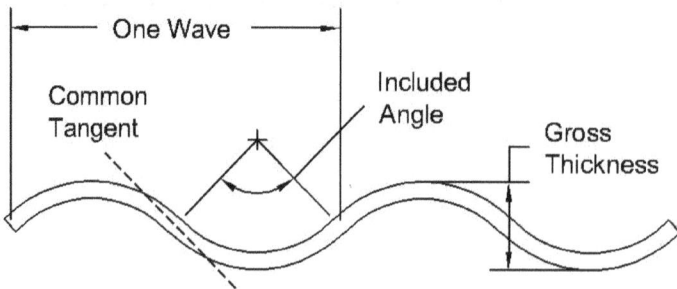

Figure 4-7. Cross section of sine wave web coupon [86, 90, 96].

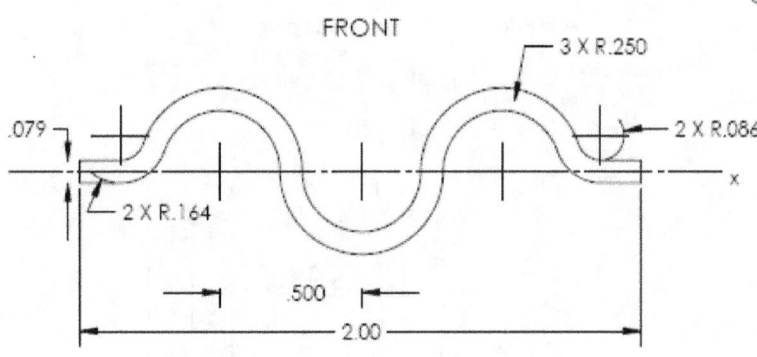

Figure 4-8. Cross section of CMH-17 corrugated coupon [90].

4.3.3.2 Flat Coupon Test Methods

Lavoie, Jackson, and their colleagues [97-99] were among the first researchers to develop a flat-coupon test fixture to evaluate different material systems, optimum laminate lay-ups, various triggers, and coupon scaling effects. The test fixture that was developed, shown in Figure 4-9, consists of a sliding top plate guided by four rods and bushings. The coupon is independently braced against global buckling by four support rods with inlaid knife edges. The fixture was designed for use with two specific coupon sizes. The larger coupon is loaded in the fixture perpendicularly to the smaller coupon, and the smaller support rods are replaced by the larger rods shown in the figure as dashed lines. For both coupon sizes, load is applied at the center of the sliding plate quasi-statically through a seated steel ball, and dynamically through a load-distributing polymer cylinder. This test fixture was later used by Johnson [95] to test flat coupons and Bolukbasi and Laananen [100] to compare the crush behavior of flat coupons with that of composite angles and channels.

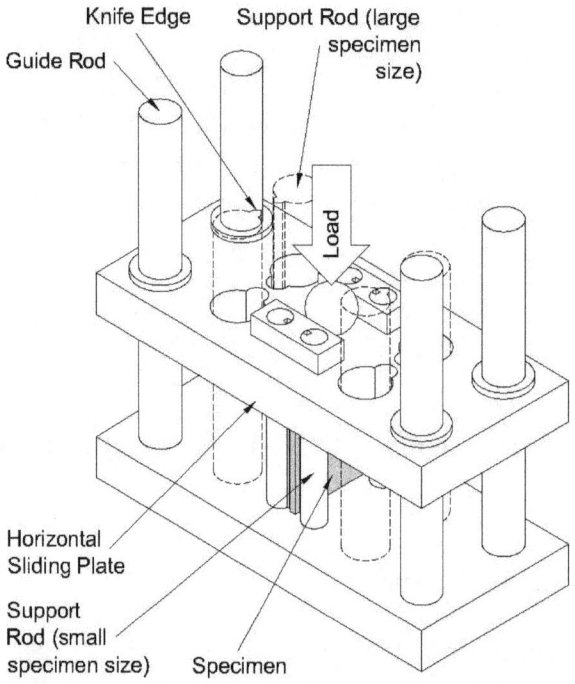

Figure 4-9. Crush test fixture design of Lavoie et al. [97-99] for flat coupons (shown configured for quasi-static loading of the smaller specimen size).

Modifications to this test fixture concept were later made by Dubey and Vizzini [101] as shown in Figure 4-10a. The modified test fixture consists of four guide rods that simultaneously support the coupon against global buckling and guide a moving block through which the quasi-static load is applied. Cauchi Svavona and Hogg [102] also modified the test fixture to include moveable knife edges that accommodated various plate widths and thicknesses as shown in Figure 4-10b. Additionally, the sliding plate was replaced with a loading block that could pass between the knife edge supports.

SUMMARY OF PROGRESS

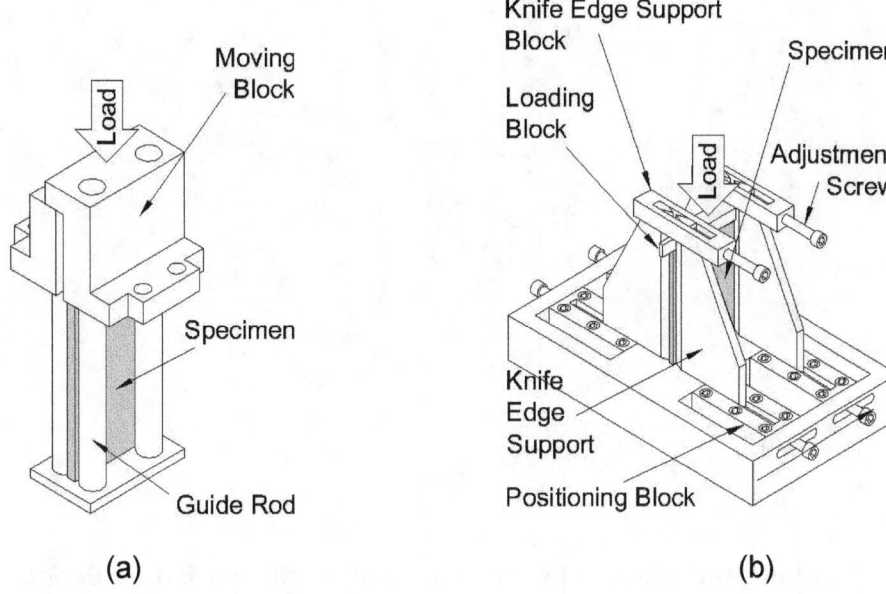

Figure 4-10. Crush test fixture designs of (a) Dubey and Vizzini [101] and (b) Cauchi Savona and Hogg [102].

A common characteristic of the three test fixture designs shown in Figures 4-9 and 4-10 is that during crushing, the specimen is forced to tear around the specimen supports. Figure 4-11 shows a representation of a crushed specimen where such tearing around the supports has occurred. Such tearing results in higher energy absorption than for a coupon that is allowed to crush without such supports and may not be representative of a structural application. Additionally, the full length supports do not allow unmitigated interlaminar crack growth in the coupon. This extra constraint may act to reduce delamination and hinder the opening and growth of cracks, also resulting in higher energy absorption than for a coupon allowed to crush unconstrained. Finally, debris may become trapped in the fixture, hindering the crushing process and leading to increased friction and binding.

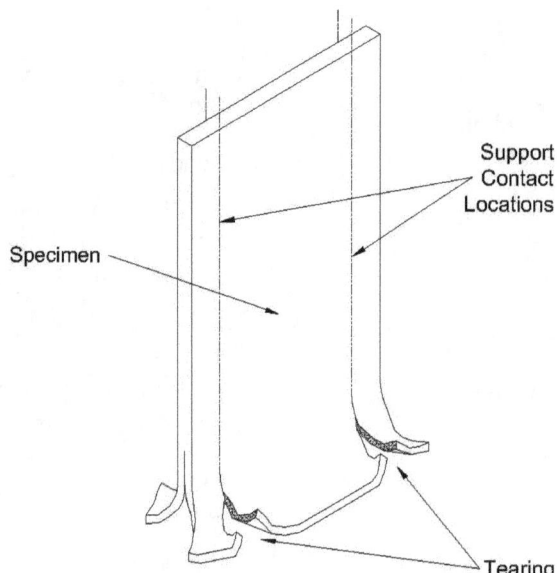

Figure 4-11. Crushing and tearing of a flat coupon using the test fixture design of Lavoie et al. [97-99]. Phantom lines represent the locations where the support rods contact the coupon and tearing occurs.

Engenuity Limited [103] has developed a series of test fixtures to crush flat coupons without coupon tearing. The most recent model is shown in Figure 4-12. The flat coupon is housed in the fixture between friction reducing Delrin sliders, and secured by a cover plate and access door. During the test, the coupon is loaded via the loading slide, supported against buckling by the surrounding housing, and crushed against the spacer block below. In the region of the crush zone, however, the coupon is unsupported over a gap height, which can be adjusted with drop-in spacer blocks of various thicknesses. The gap provides a passage for the crushed, splayed laminate and debris to escape without tearing or interfering with the test in progress. The main body of the fixture can be tilted back via the hinge in the fixed base, which allows for convenient changing of the coupon and adjustment of the spacer blocks.

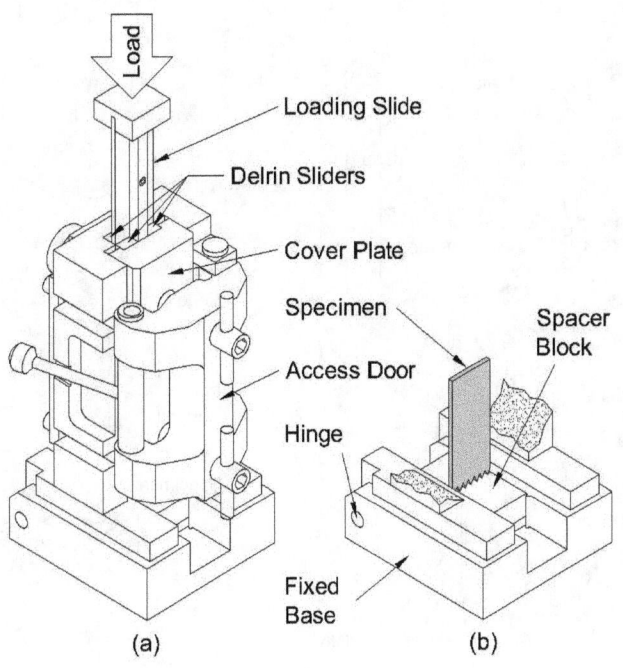

Figure 4-12. Test fixture developed by Engenuity Limited [103]: a) Full test fixture, and b) Cut-away view revealing the specimen and the spacer block.

Other test fixtures have been designed with unsupported gaps for crushing flat composite coupons without inducing tearing. The fixture developed by Takashima et al. [104] provides a lower support to the specimen using a matched-thickness steel plate as shown in Figure 4-13a. The top of the specimen is crushed against the upper loading plate. The fixture of Feraboli [105] (Figure 4-13b) employs coupon supports that pass through cut outs in the sliding top loading plate. The reduced sections of the top loading plate that contact the specimen are not adjustable, a potential problem for coupons whose thickness is different than the width of the reduced plate sections.

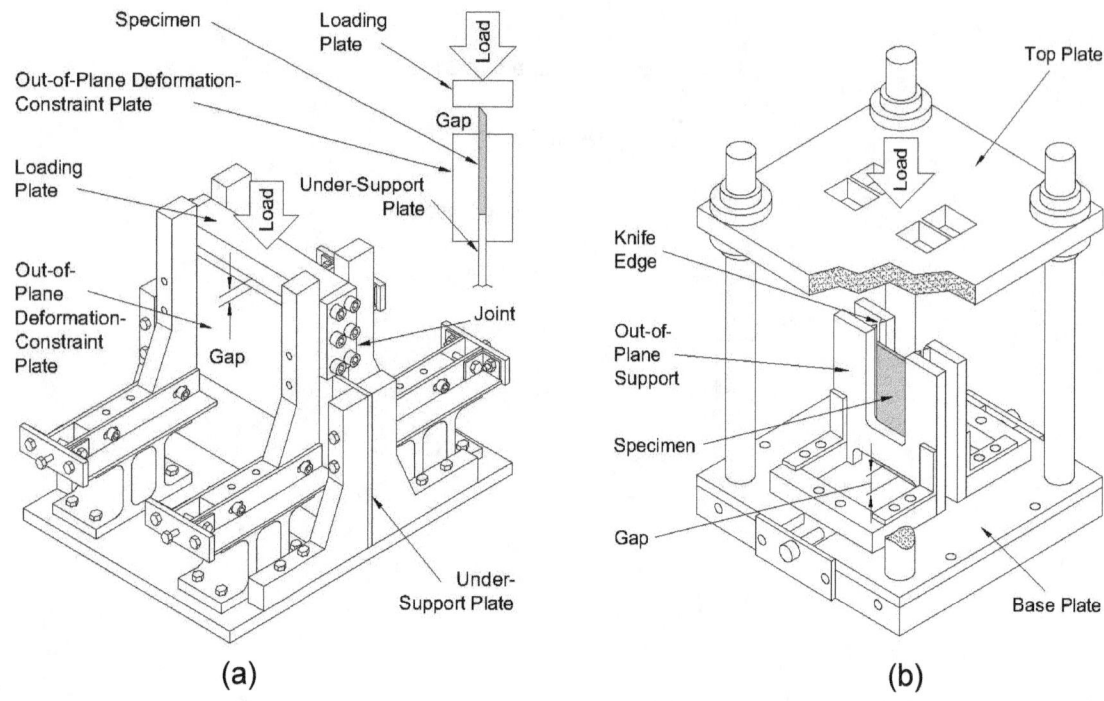

Figure 4-13. Flat coupon crush test fixture designs of: (a) Takashima et al. [104], and (b) Feraboli [105].

Following the review and evaluation of previously developed flat coupon crush test methodologies, Garner and Adams [91] developed a flat coupon crush test fixture similar in function to that developed earlier at Engenuity Limited [103]. As shown in Figure 4-14, this fixture accommodates variable coupon thicknesses using adjustment screws in the buckling support and the gap height is varied using spacer blocks of various thicknesses. The fixture design allows for both a front and side viewing and high-speed video recording of the coupon's crush zone during testing to allow observations of the failure mode(s) and crush behavior. While used initially for quasi-static testing, the fixture has also been used with drop tower testing equipment.

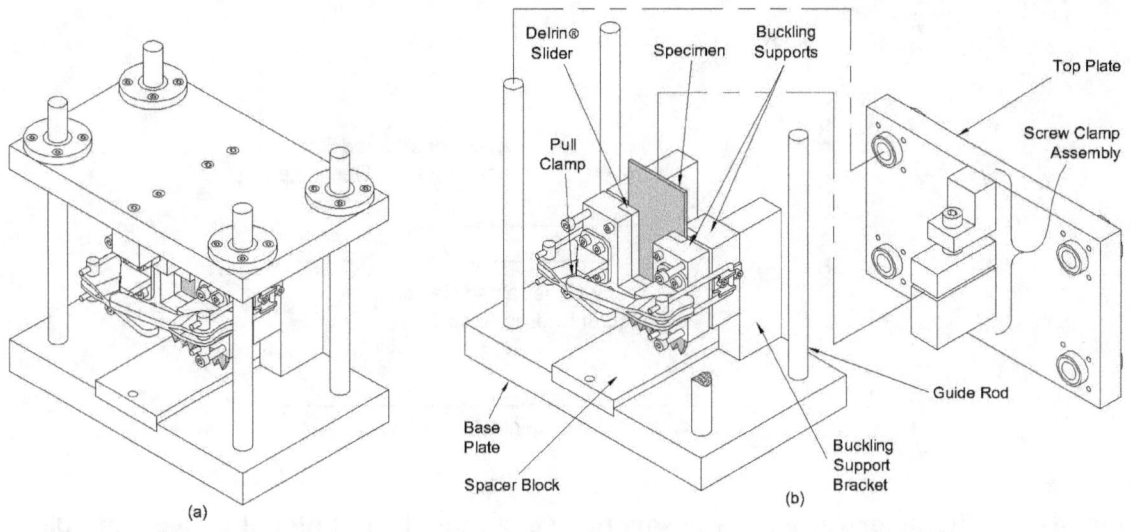

Figure 4-14. Flat coupon test fixture of Garner and Adams [91]. (a) fully assembled, (b) top plate removed.

The four test fixtures shown in Figures 4-12 through 4-14 have overcome coupon tearing (Figure 4-11) by incorporating an unsupported region, or gap, wherein the coupon is potentially allowed to crush without artificial constraint. To ensure free yet supported crushing, specification of the proper gap height is very important. If too large, the gap may allow buckling and catastrophic failure to occur. If too small, the buckling supports may overly constrain the coupon and not produce proper crushing. A characteristic plot of the energy absorption versus gap height obtained from flat coupon crush testing is shown in Figure 4-15. Similar to the determination of the proper gage length to be used for composite compression testing, the proper gap height is dependent on the stiffness, strength, and thickness of the coupon to be tested. Roberts and Barnes [73] have reported that the appropriate gap height may be determined from a limited number of test performed using different gap heights.

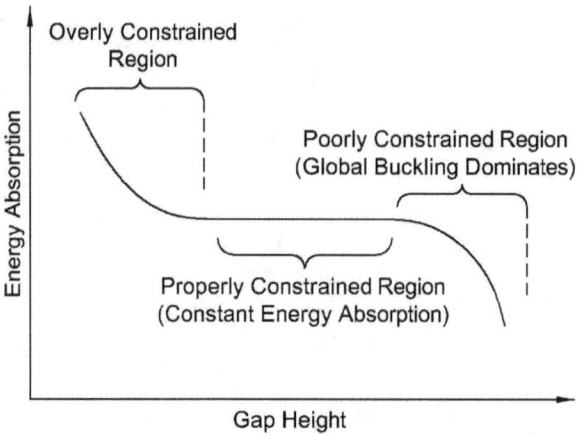

Figure 4-15. Characteristic energy absorption versus gap height plot obtained from flat-coupon crush testing.

4.3.4 Element-Level Test Methods

Simple composite structures, such as tubes, have been crush tested in an effort to help characterize the crush behavior and energy absorption of more complex structures. Such tubes, characterized as an element-level test specimen, are commonly used in the automotive industry to characterize energy absorption because of their similarity in configuration with automotive energy absorbing structures such as the upper and lower front rails. Although element-level test methods are more focused towards an intended application than coupon-level test methods, they are still considered to be part of the first level, or base, of the Building Block approach (Chapter 1) as they are typically used to assess material behavior.

If designed properly, such tubes may fail in a stable progressive crush mode. Figure 4-16 shows a typical test setup used for crush testing of composite tubes. Since tubes are self supporting structures, generally no specialized test fixture is required. The upper end of the tube is contacted directly by the moving upper platen and an external plug trigger (if used) is placed on the lower platen. As discussed previously, this trigger is used to promote progressive crushing and preempt catastrophic failure.

Among the first publications regarding crush testing of composite tube specimens was Thornton and Edwards [14] in the late 1970's. Such early studies identified that composite tubes provide a combination of high energy absorption and low weight, making composites an attractive candidate for primary energy absorbing automotive structures.

Although several cross sectional shapes have been considered for element level testing, the most commonly investigated shapes have been the circular and rectangular tube. Of these two cross sectional shapes, the circular tube has been found to produce the largest value of SEA [106-108]. Hull [109] made extensive use of circular tubes for observing the failure modes and mechanisms of crushing. Fairfull and Hull [89] used circular tubes to study frictional effects during crushing. In a separate test, they were able to determine the coefficients of friction by rotating the ends of the circular coupons against platens of various surface textures.

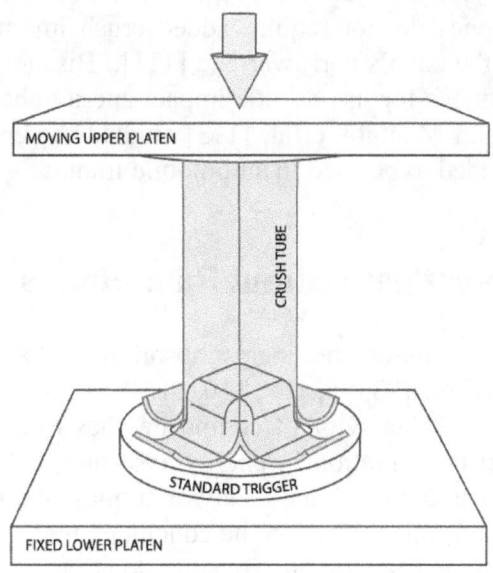

Figure 4-16. Schematic of a tube crush test with internal plug trigger.

Square and rectangular cross section tubes are also frequently used in crush testing as they represent common structural shapes, especially in automotive frames. Brimhall [93] used square tubes to quantify the contribution of friction to the total absorbed energy. Thornton [85] and Czaplicki et al. [84] used square (and circular) tubes as a platform for comparing the effect of crush initiating triggers on energy absorption.

The shape of the composite tubes used in element-level crush testing has been found to have a significant effect on energy absorption. Hull [109] noted from his experiments that square and rectangular tubes did not generate the conventional crush zone morphology in their flat-walled sections, which tended to fail by buckling. He concluded, therefore, that square and rectangular tubes were less energy absorbent and structurally weaker cross sections when compared to circular ones. Caruthers et al. [110] reported that square and rectangular tubes are less energy absorbent than circular tubes due to the stress concentrations of the corners. Jacob et al. [111] ranked common cross sections in order of decreasing energy absorption capability: circular, square, and rectangular.

Other geometric factors involving tube crush testing have been investigated. Dubey and Vizzini [101] investigated the effects of tube diameter to thickness (D/t) ratio and found that the SEA measured during crush testing increased with increasing values of the D/t ratio. Elgalai et al.

[112] found that for tube crush tests performed with several length-to-diameter (L/D) ratios, the highest value of SEA was obtained for a L/D ratio of approximately 5.

Other element-level specimen configurations have been investigated. Conical shells of circular and square cross section are yet other geometries employed in crashworthiness studies [113]. It is interesting to note that cones do not require added crush initiating triggers because stable crushing naturally begins at the cone's narrower end [111]. Bisagni et al. [114] investigated the behavior of circular conical tubes for use as side impact energy absorbers in Formula One race cars. Johnson et al. [115] and Mamalis et al. [113] studied the energy absorbing behavior of hourglass cross sections intended to be used in automobile frames.

4.3.5 Testing to Investigate Strain Rate Effects

The effect of load and strain rates on the energy absorbing behavior of composites has been investigated by several researchers [93, 95, 109, 110, 116-120]. Attempts to generalize the effect of load and strain rates on crush behavior of composites has to date been inconclusive [110]. Brimhall [93] concluded that the variation in energy absorbing behavior at different load rates was due largely to the change in the frictional behavior at quasi-static versus dynamic load rates. Through an experiment that minimized friction, he concluded that the specific energy absorption was virtually the same at both quasi-static and dynamic load rates. However, Jacob et al. [116] states that the load-displacement curve, initial peak load, magnitude of the energy absorbed, and the time required to absorb this energy are all functions of the crushing speed. Hull [109] observed that some fiber arrangements are affected by load rate and have an associated change in crush mode. Jacob et al. [119] reported that the strain rate can affect the matrix behavior and the failure modes, and concluded that beyond a certain threshold velocity, the composite material's energy absorption capacity suddenly drops.

According to Jacob et al. [111], the energy absorbing mechanisms vary with load rate. The important factors for energy absorption at high load rates were found to include the magnitude of the energy dissipated in delamination (interlaminar crack growth), debonding, and fiber pull-out. For low load rates, the important factors were the strain energy absorption of the fibers and the geometric configuration.

Farley and Jones [120] suggested that if the mechanical properties that control the failure mechanisms are influenced by the strain rate, then the crushing speed is likely to affect the energy absorption behavior of the specimen. For example, the matrix stiffness and failure strain may be functions of the strain rate, so it is expected that the energy absorbed through crack growth during transverse shearing or lamina bending failures will be a functions of the crushing speed. Conversely, only transverse shearing is exhibited in brittle fiber reinforcements whose mechanical properties are generally insensitive to strain rate. The fracturing of the lamina bundles is generally not a function of crushing speed. However, the coefficient of friction can be a function of speed and therefore its contribution to the energy absorption during the lamina bending failure mode is expected to depend on the crushing speed.

Although there are currently no standard test methods to investigate strain rate sensitivity in composites, the effect of strain rate has been investigated in other materials as well as in the fracturing of adhesively bonded composite joints. For metallics, considerable research has been performed to investigate strain rate effects of metallic sheets in tension (see, for example [121, 122]). Although no test standard currently exists, the International Iron and Steel Institute (IISI) has developed and published recommendations for dynamic tensile testing of sheet steels [123]. As a sign of both the significance and maturity of the field, a reference book on strain rate testing of metallic materials that is focused on usage for automotive crash modeling has recently been published [124].

For plastics, research focusing on strain effects of plastic specimens in tension has lead to both an ISO test standard in 2007 [125] and a SAE recommended practice in 2008 for high strain rate tensile testing [126]. Research has also focused on rate effects in the fracture of automotive adhesives. Under funding from the Automotive Composites Consortium, Dillard and colleagues [127,128] developed high rate tests to evaluate the dynamic fracture properties of commercial epoxy adhesives.

Note that for both metallics and plastics, dynamic tensile testing is used to investigate the strain rate effects of these materials for use in the automotive industry, where crash modeling involves compression loading. For composites, it is well understood that failure mechanisms are different under tensile and compressive loading. As a result, there is little confidence that high strain rate tensile testing of composites will be useful for investigating strain rate effects in crash scenarios.

4.3.6 Summary: Crashworthiness Test Methods

Table 4-2 summarizes the current types of test methods available for assessing the energy absorption capabilities of composites. Included in the summary table are the primary advantages and disadvantages as well as their relevance to PCIVs. Currently there are no standardized tests by which crashworthiness may be assessed. Recommendations concerning the development of standardized test methods for composite crashworthiness are presented in Chapter 5.

Table 4-2. Summary of Crashworthiness Test Methods.

Type of Test Method	Advantages	Disadvantages	Relevance to PCIVs
Corrugated coupon testing	No test fixture required	Requires special fabrication of shaped specimen	Delamination suppressed SEA values In use for CMH-17 Round Robin
Flat coupon, supported gage section	Small, inexpensive specimens	Requires a specialized test fixture Specimen forced to tear around supports	Failures produced not representative Several methods developed
Flat coupon, unsupported gage section	Small, inexpensive specimens Laminate crushing without artificial constraints	Requires a specialized test fixture	Crashworthiness properties of laminates (Level 1 of Building Block) In commercial use for material screening and characterization
Tube testing	Representative of structure	Larger and more expensive specimen Results dependent on tube shape and geometry	Crashworthiness properties of laminates (Level 1 of Building Block) Modeling validation

4.4 Summary of Progress: Composite Crashworthiness Modeling

Crash analysis has become a pivotal instrument in the development of new vehicles, whether PCIV or conventional. In addition to the marketability of new vehicles relying on high scores in the industry accepted safety tests, regulatory requirements should also not be underestimated. The extent to which the large automakers rely on predictive crash analysis is exemplified by the fact that 70% of Chrysler Group simulation activities in 2002 were crash related [129]. The variety of load cases for crashworthiness modeling is extensive, as illustrated by the listing of load cases presented by Gohlami et al. [130] and summarized in Table 4-3. In order to move PCIVs to high volume production, automakers will need confidence in the crash predictions to a level comparable with that currently in place for conventional metallic structures. Without crash simulations, contemporary new car development projects would not merely be inconvenienced, but would not be possible [131].

Table 4-3. Crashworthiness Load Cases Considered For Automotive Development [130].

Crashworthiness Load Cases

Whole Car	Components
Frontal Crash	Roof Crush
Rear Crash	Bumper Testing
Side Crash	Door Intrusion
Pole Test	Baggage Restrain
Rollover	Rollover Protection Systems
Compatibility	Pendulum Test ECE R21
	Head Impact FMVSS 201
	Seat Pull Test ECE R14

The ability to perform computational simulations of a vehicle crash involving composites has been a goal of the automotive industry for at least twenty years. As composites have become more of a mainstream material choice in other industries, particularly the aerospace, recreation, and commercial sectors, they have been recognized as excellent material choices, particularly for weight-critical applications. In these market sectors, however, the design of composite components generally requires only the design to an initial failure. At most, such analyses include only the initial failure and not its progression in a composite material, either through matrix-damage (such as microcracking) or fiber failure. In most market sectors, this first fiber failure signifies the point of failure in the composite structure, and the termination of a computational simulation. In reality, the composite structure still has a considerable degree of load carrying capacity. In contrast, the first fiber failure often represents the starting point of a composite crash analysis, which requires the simulation of a composite structure through a progression of crushing, in which the cumulative energy absorption must be predicted.

As a result of the difficulties associated with crash modeling of composites, crash component testing has remained as the primary means by which the crush behavior and energy absorption of composites are evaluated. However, considerable progress has been made towards the development of computational modeling approaches for crash modeling of composite structures. This section presents an overview of composite crashworthiness modeling methodologies and summarizes progress in the development of predictive capabilities.

4.4.1 Overview of Crashworthiness Modeling

Virtually all composite crashworthiness modeling efforts to date have included the use of an existing conventional, explicit finite element code that has been used previously in the crash simulation of conventional metallic structures. As such, these modeling efforts generally begin with preexisting capabilities for crash modeling of metallic structures, including: initiating the crash event, modeling the contact of impacting bodies, modeling material yield, post yield plasticity, and possible element deletion after a failure criterion has been reached. As a result, a major developmental aspect of crashworthiness modeling for composites is the incorporation of failure criteria and damage development models for composite materials. A second aspect is the development of a methodology for modeling crushing of a composite material along the "crush front" that is in contact with the impactor.

An overview of crashworthiness modeling may be presented by considering the simulation of a crush experiment of a rectangular column of a composite material as shown in Figure 4-17. The column is impacted by a mass moving at an initial prescribed velocity. Upon contact, the column begins to fail at the point of impact, and develops a crush front that propagates along the length of the column. Typically, the initial load peak is larger than the sustained crush load, as shown by the schematic load versus displacement plot in Figure 4-17. The load remains relatively constant during the process of progressive crushing. At some point, the crush process terminates when the initial impact energy has been completely expended.

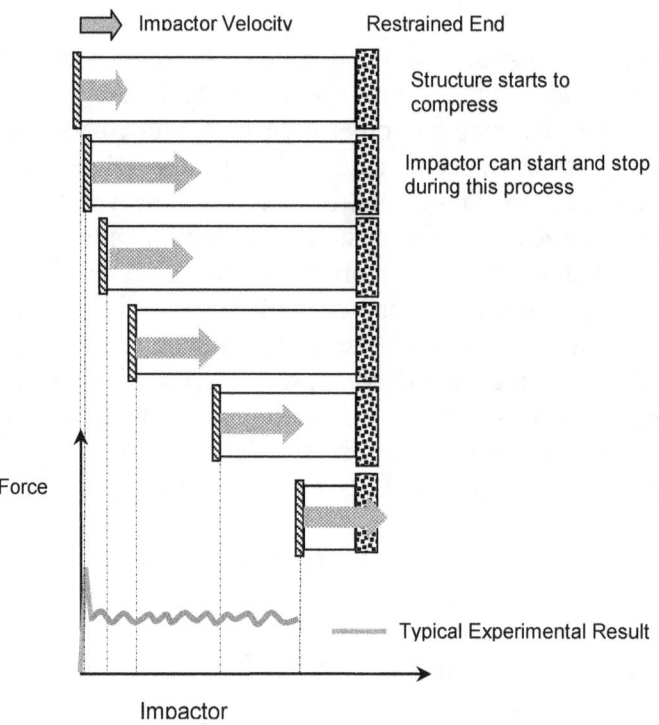

Figure 4-17 Schematic representation of a crush experiment performed on a composite material.

4.4.2 Categories of Crush Front Modeling

The crush phenomenon in composites is widely accepted as involving different failure modes than those observed in conventional metallic materials. In many ways, the term "crush" when applied to composites is used to define the complex and largely unexplained mechanisms of energy absorption, some of which take place on micro or sub-micro scales. As a result, researchers have addressed the problem of crush zone modeling with different methodologies and at different length scales. In general, crush front modeling efforts to date can be classified into four categories: progressive damage modeling, continuum damage modeling, multi-scale modeling, and phenomenological modeling. A brief description and summary of progress is provided for each category of crush front modeling in the following sections.

4.4.2.1 Progressive Damage Modeling

The progressive damage models used for composite materials have their roots in the procedures applied to predict failure using Classical Laminate Theory to predict

1. Damage Initiation
2. Damage Progression
3. Final Failure.

This approach was then extended and applied to the linear, static finite element analysis process in order to more expediently model complex laminates and identify the damage initiation through first-ply failure by means of one of the many established failure theories for either a fiber or matrix failure. In the linear models, damage progression is usually affected by manual modification of the laminate definition. This process is performed on an element-by-element, ply-by-ply basis to take into account the complex interactive nature of damage and determination of which ply (and in what direction) to degrade based on the previous failure condition. This process includes additional matrix microcracking, fiber breakage and pull-out, delamination between layers, and crack propagation failures on the element scale. As the composite laminate is often multidirectional, a failure in one direction will often destroy the load carrying capability in other directions. As a result, it is common to achieve the degradation on a layer-by-layer basis. In this way the progressive failure model with linear finite element analysis can be used to estimate the damage progression and strength of an entire structure [132]. The process is repeated after each iteration of the analysis until either the load paths are redistributed to prevent further damage growth or the damage continues and element failure leads to further damage and failure of the structure.

The development of the progressive damage model for use in explicit finite element analysis codes was initially deployed in an automated manner. However, it was based on a similar approach to that adopted in the linear analysis case. For this reason, it was used for failure prediction of the back-up structure without any consideration for application within the crush front.

Consider next the general procedure employed to perform a computational simulation of the crush experiment shown in Figure 4-17 using a conventional, explicit finite element code and a progressive damage/continuum damage model. As illustrated schematically in Figure 4-18, the column is discretized into a single row of rectangular finite elements. The modeling approach shown illustrates a common progressive damage modeling procedure used with LS-Dyna [133], a commercial finite element code commonly used for crashworthiness modeling. The crash analysis begins with the impactor establishing contact and the elements of the composite structure starting to load in prescribed time increments. As the stresses in a ply continue to increase to a point where they exceed the designated failure criteria, the stiffness properties in the failed ply are degraded to zero over a fixed number of time steps [134]. Optionally, in some implementations such as MAT8 in MSC.Dytran, material properties are degraded in complementary directions. The analysis continues monitoring the failure criteria for the remaining intact directions within individual plies. When they exceed the failure criteria in operation, they too are degraded. At some point, the state of damage in the element reaches a point where element failure is determined according to a prescribed condition, which could be an arbitrary minimum solution time step or the fact that all plies have been degraded to zero stiffness. At this point, the failed element is deleted from the mesh, creating a new "gap" between the impactor and the structure. The impactor continues to move towards the structure, contact reinitiates, and the structure begins to reload.

In order to produce failure at the "crush front", a crush parameter (referred to as "SOFT" in LS-Dyna) is used to reduce the effective element failure allowables in elements adjacent to those which have experienced failure. Elements adjacent to those which have been deleted become "crush-front" elements, with reduced strength allowables being employed within the failure criteria. This process continues, creating the force versus impactor displacement plot shown in Figure 4-18.

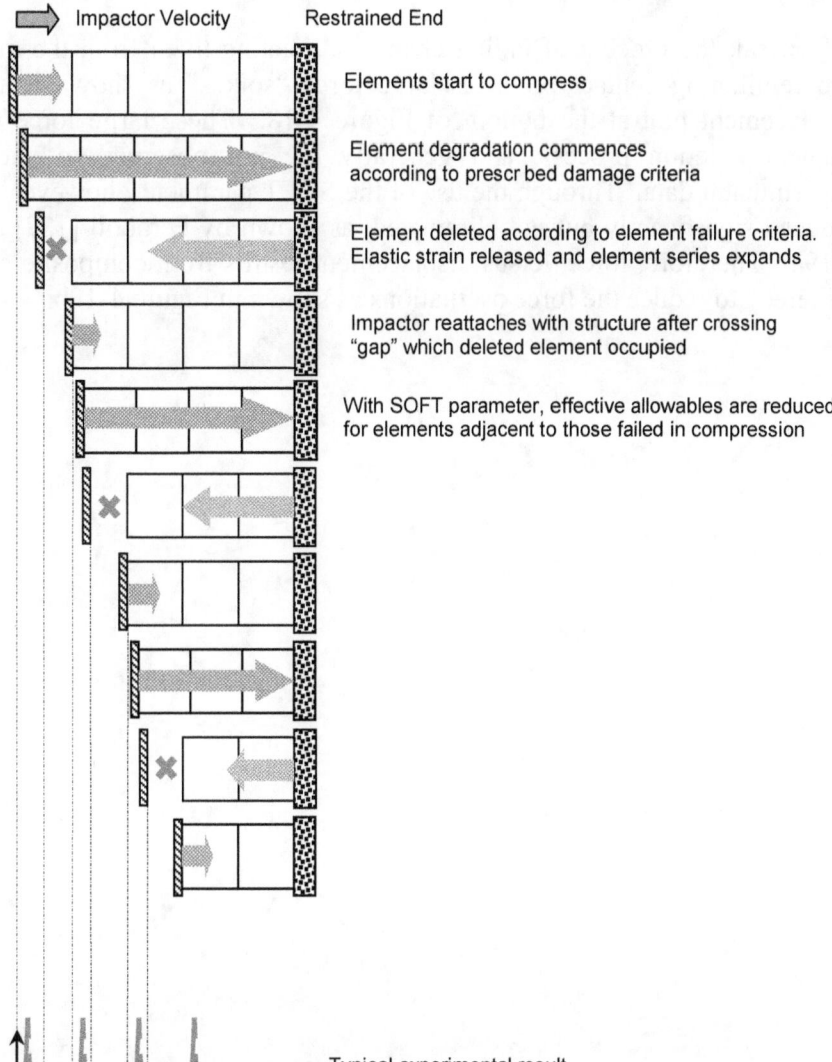

Figure 4-18. Force ...tional explicit finite ele... ...crush experiment.

In LS-Dyna, MAT 54 uses the Chang-Chang [135] and MAT 55 uses the Tsai Wu failure criterion [136]. Both use an enhanced damage material model that uses the SOFT parameter as a reduction factor for the material strength in the crush front elements. In cases where component crush testing is available in advance of the simulation, it is possible to obtain a value for the SOFT parameter using a "trial and error" approach. A filtered force versus displacement plot (similar to that shown in Figure 4-19b) is compared to that obtained from crush testing. Other tuning factors, such as the contact stiffness curve, can also be varied to influence the load versus displacement response obtained from simulation. However, the SOFT parameter is not believed to be related to any physical or measurable quantity. Further, the response obtained using the SOFT parameter is mesh size dependent, as highlighted by Pinho et al [134]:

> "To model failure, the approaches described above suffer from a severe mesh dependency problem related to strain localization during the fracture process."

In general, the process of failing elements, deleting them from the mesh and creating a "gap," and reinitiating contact produces large force "spikes" as shown in the force versus impactor displacement plot at the bottom of Figure 4-18. These large force spikes are artifacts of the element deletion process, and generally are not observed to such a large magnitude in experimental data. Through the use of the SOFT parameter, however, these force spikes may be reduced significantly but not eliminated, as shown by Feraboli [137] and summarized in Figure 4-19a. Therefore, force versus displacement results from composite crash simulations are often "filtered" to reduce the force oscillations as shown in Figure 4-19b.

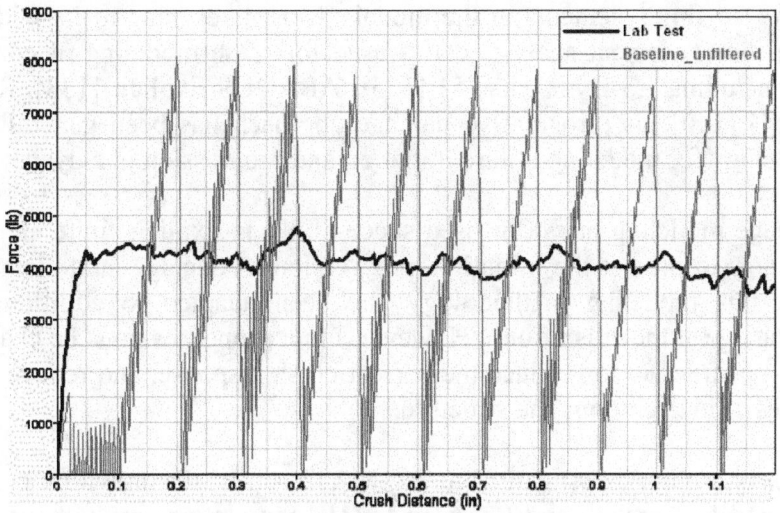
a. Unfiltered simulation results.

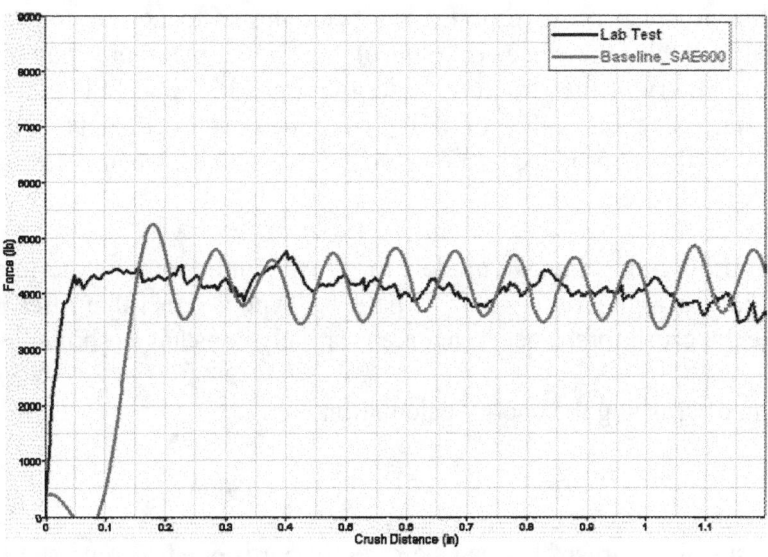
b. Filtered simulation results (600 Hz).

Figure 4-19. Effects of filtering results from composite crash simulations [137].

Although post-analysis filtering may be used to reduce these load peaks when presenting results, such peaks are computed at the frequency that the mesh size and failure position dictate, and are introduced into the structure. Such load peaks, artifacts of the modeling method used at the crush front, may lead to a premature prediction of failure resulting away from the crush front in the back-up structure.

Although the use of progressive damage modeling does not explicitly model the generation of damage within the composite structure, it accounts for damage through the reduction of stiffness properties of affected finite elements in the model. Most likely due to its wide availability in most commercial finite element analysis codes used for crash modeling of conventional metal body structures including LS-Dyna MAT54 [133], ABAQUS/Explicit [138], RADIOSS [139], and PAM-CRASH [140], the progressive damage approach has been generally accepted as a practical methodology for modeling failure initiation and progression in a dynamic event.

Progressive damage modeling has been used successfully in explicit finite element analysis of composite structures to predict the initiation, progression of damage, and element failure. For crush modeling, this modeling methodology has been adapted to additionally predict the progression of damage at the crush front. However, input properties must be changed from those used in the back-up structure to produce the correct crush response, and reduce the elevated but still present values of load entering the structure.

Some of the more thorough investigations using MAT54 in LS-Dyna with the SOFT parameter have been presented by Feraboli and Rassaian [141]. Simulation results have shown that it is possible to generate a wide range of force versus displacement responses by varying only the SOFT parameter. The value of the SOFT parameter selected to best produce the experimentally-obtained force versus displacement response has been shown to not be constant for a specific composite material. Rather, the selected value of the SOFT parameter has been shown to be dependent on the geometry of the structure undergoing crush. Xiao [142] has confirmed these results, showing that using the value of SOFT obtained from crush testing of a sinusoidal specimen (Figure 4-8) did not produce acceptable force versus displacement results when applied to other cross sectional shapes.

Relating the use of progressive damage models for crush simulation to experiment results, it has been established that for a given section in Type 1 or Type 2 crush that after initiating a crush front, the experimental crush forces generated are relatively constant as shown in Figure 4-20.

Over a given length of crushing the energy calculation is

$$(Energy) = (Force) \times (Crush\ Distance).$$

In the instance of sustained crush this can be directly partitioned to represent the intermediate length of a finite element;

$$(Energy) = (Force) \times (Element\ Length) \times 100\%$$

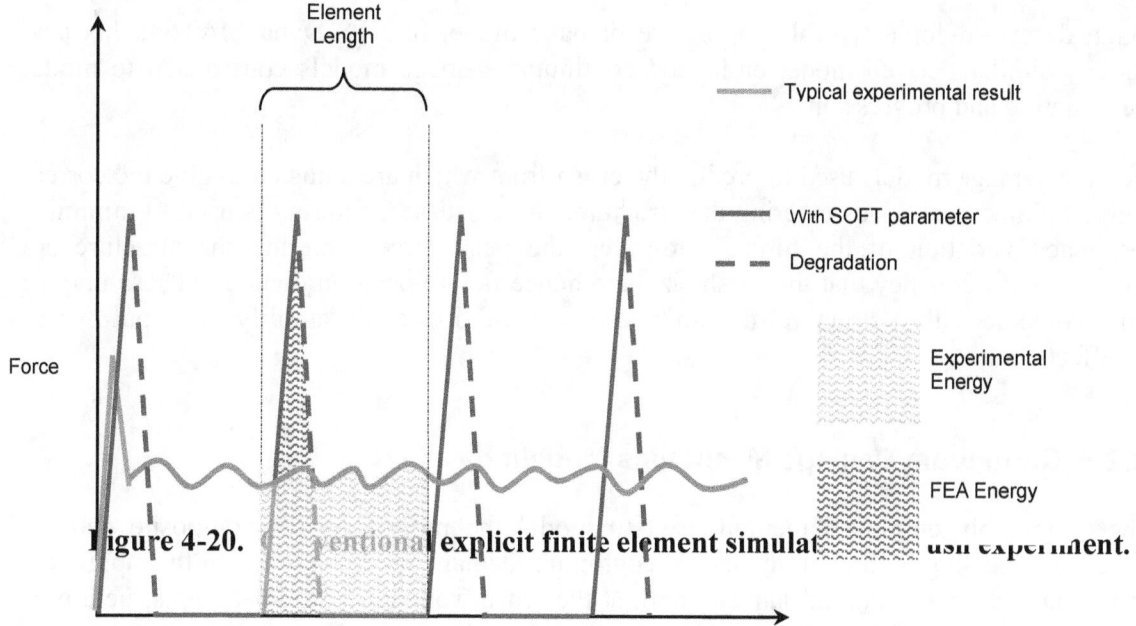

Figure 4-20. Conventional explicit finite element simulation vs experiment.

In the LS-Dyna MAT 54 progressive damage model, crushing of a "crash front" element (i.e. an element with SOFT-modified failure properties), begins when the impactor contacts the element. The element begins to load and strain according to the un-degraded laminate stiffness to a point where the Chang-Chang failure criterion [135] signifies initial damage (typically <2% strain in compression), at which point the plies degrade. Depending on the specification used, either an instant degradation to zero stiffness occurs in the ply, or degradation over a fixed number of time steps to zero may be implemented.

To balance the energy (area under the graph) with experiment, the mean force would need to be increased proportionally with the reduction in crush length from 100% to 1.5%,

$$(Energy) = 50 \times (Force) \times (Element\ Length) \times 2\%$$

This large force is unsustainable and therefore the strain-to-failure is increased to non-physical levels in compression such as 40%.

Overlooking the non-physical and non-predictive capability in the back-up structure calculation that result, consider the energy absorbed in the finite element simulation with the artificial 40% compressive strain in all plies at element failure;

$$(Energy) = 2.50 \times (Force) \times (Element\ Length) \times 40\%$$

In this case the mean crush force of 40% of the element length needs to be 2.5 times higher than the experimental results in order to maintain the correct energy levels. This result demonstrates the inability to predict failure behind the crash front, an essential requirement for predicting failure in the back-up structure. This usually causes significant model instability and element ill conditioning.

Although described for a typical progressive damage model like LS-Dyna MAT54, this also affects any similar derived model or indeed continuum damage models constructed to model failure initiation and progression.

Progressive damage models used to predict the crush front which are adjusted to give the correct SEA give inflated peak loads entering the structure. As discussed, filtering is used to minimize the presented variation of the force. However, the peak forces entering the structure are computed at the frequency that the mesh size and hence failure position dictate. These may or may not coincide with a structural dynamic response and artificially amplify the input due to modal effects.

4.4.2.2 Continuum Damage Mechanics Modeling

Another commonly employed method used to model the progression of composite damage during crash modeling is through the use of continuum damage models [143]. In this approach, the initiation and progression of damage, both at the crush front and in the back-up structure, is handled at a macroscopic level in each finite element using variables referred to as damage parameters. Each damage parameter accounts for the state of a particular type of damage within the element, and the material stiffness property degradations associated with each parameter are intended to account for this type of damage within the element.

Generally, the value assigned to each damage parameter is based on the state of stress or strain within an element, and unlike progressive damage models this approach does not usually use failure criteria to establish the initiation of failure as there is a predefined stiffness response for all strain conditions. Element failure is defined to occur when one or more of the damage parameters reach a critical value. An exception is the Radioss Material Law 25 [144] which uses a hybrid approach to initiate failure using a failure criterion but to use a plasticity law thereafter.

Although the use of continuum damage modeling does not explicitly model the generation of damage within the composite structure, it accounts for damage through the reduction of stiffness properties of affected finite elements in the model. Most likely due to its wide availability in most commercial finite element analysis codes used for crash modeling of conventional metal body structures, including LS-Dyna MAT58 [133] and RADIOSS [139], the continuum damage approach has been generally accepted as a practical methodology for crash modeling.

Warnings have been expressed concerning the applicability of progressive and continuum models, and the LS-DYNA MAT55 and MAT58 material models in particular, towards modeling high compressive strains implicit in crashworthiness simulations. Schweizerhof et al. [145] in 1998 stated:

> "...*Thus any analysis assuming a 2D continuum model – such as in shell analysis – have almost no reasoning for this regime... in particular, the regime of applications has to be*

kept into the limits of continuum mechanics. In general, analysis involving large strains cannot be performed unless further assumptions are taken into account."

This was reinforced by Pinho et al. [134]:
"...Strictly, problems in the areas of crashworthiness or high-energy impact, should thus fall out of the scope of most CDM-based FE analyses, since energy absorption is the main motivation for performing the modelling; these analyses should proceed no further than damage initiation".

As with other modeling approaches, the use of the continuum damage modeling is subject to particular limitations, and must be validated using appropriate test data. The degree to which this modeling approach, when based on a specific material and validated using a specific test condition, may be applied to other materials and loading conditions for crash modeling remains an active area of research. The continuum damage model is also afflicted by the same issues regarding the incompatibility of peak force and energy levels when considering compressive failure below 100% crush length.

4.4.2.3 Multi-Scale Modeling

Fundamentally, the multi-scale modeling approach involves determining the damage evolution and failure of finite elements within a structural model by use of analyses performed on a refined model of a reduced length scale. Essentially the output from the reduced length scale model becomes the specification for a continuum damage model for the structural-scale simulation. To this effect, multi-scale modeling is afflicted by all the limitations fundamental to the continuum damage modeling discussed previously.

For composite crash modeling, the multi-scale modeling approach has been used to model composite tubes made of braided composites [146-148]. The reduced length scale model was selected as a unit cell of the braided composite architecture, composed of the smallest repeating element of the braid architecture for a single braided layer. For crash modeling, this approach has been used in two research investigations supported by the Automotive Composites Consortium (ACC). Flesher and Chang [146, 147] developed a multi-scale modeling approach to predict the crush behavior of braided composite tubes using ABAQUS/Explicit. A VUMAT user subroutine was developed to obtain the material response based on a unit cell model of the braided composite architecture. The unit cell was composed of nine subunits consisting of braider tow, axial tow, or matrix material. The response of the unit cell to the states of stress in each structural element stresses was calculated, and failure of the tows and matrix material was based on localized stresses calculated from the unit cell model. Strain rate effects were included in the model using a viscoplastic constitutive law, which was applied to the transverse and shear response of the fiber tows in the unit cell model. This modeling approach was used to predict the crush behavior of both round and square tubes with braider yarn angles of 30, 45, and 60 degrees relative to the axis of the tube. In general, model predictions were in agreement with test results, which led the authors to conclude that this modeling approach could be used to predict the energy absorption in braided composite materials.

Following the braided tube model development project by Flesher and Wang, a second multi-scale modeling research project on braided tubes was initiated by the ACC. Fish and Yuan [148] have developed a Multiscale Design System (MDS) that has been demonstrated with a subroutine within the ABAQUS/Explicit finite element code. At the refined length scale, a braided composite unit cell is meshed in significant detail, including the geometry of the braider yarns and the matrix pockets between yarns. Analyses have been performed using this unit cell model for a number of static loading cases for which coupon-level data is available and model correlations may be performed.

The published works using these multi-scale modeling approaches have only been applied to tube tests with plug-type triggers. As described previously, the use of the plug trigger does not produce Type 1 or Type 2 crush behavior. In contrast, strips of material or "fronds" are forced around the radius of the plug, causing bending failure in a predetermined manner. Thus it is not clear how these modeling approaches would perform in a simulation involving the more efficient energy absorbing Type 1 or Type 2 crush behavior.

Extending the multi-scale concept, a multi-scale multi-physics example such as the progressive compressive failure implementation in CODAM [149] falls into the same category. A Representative Volume Element is modeled using an "Analog" representation using effectively Multi-Body System techniques to account for the internal failure debris or "rubble" that can coalesce in compression, but effectively be inactive in tension. This approach effectively computes the damage mechanics laws based on a RVE by using discrete tension and compression cycles based on fuses, springs, sliders, rigid and gap representations to simulate the independent damage attributes of the fiber and matrix and their interaction with the adjacent material. The derived maximum compressive strain at total damage response for the RVE with the braided material is 15%, and for a T300 woven material in an epoxy matrix this occurs at around 5% strain. In the axial crush examples, the filtered response is comparable with the tested components. This model is afflicted in the same manner as the general continuum damage elements and should be subject to the warning and limitations referred to above for this type of modeling.

4.4.2.4 Phenomenological Crush Modeling

A fourth method of crush front modeling is referred to here as the phenomenological crush modeling method. In this method, the modeling of crushing of the composite is handled by the use of an input property referred to as the crush stress. This property, determined from coupon-based testing, provides a compression stress level at which elements at the crush front will be subjected to full-length crushing. Other than the experimentally-determined crush stress, no other model parameters are required for predicting composite crush behavior in the crush zone. Outside of the crush zone, failure of the composite may be modeled using any existing composite damage model utilizing the state of stress or strain in the element.

Currently, the only commercially available code to include an phenomenological crush model for composites is CZone for ABAQUS [150], a licensed add-on for ABAQUS/Explicit [138]. The CZone for ABAQUS product is based on patented technology developed by Engenuity Limited

[151] and implements the principal of a coupon (or component) derived crushing stress, applied directly as forces on the crushing elements for the whole extent of the element in the crush front. This same principal was demonstrated in ESI PAM-CRASH and referred to as the Energy Absorbing Contact (EAC) [152]. In CZone, the implementation of these crushing forces is handled by a combination of the crushing contact definition and the crushing material properties which are associated to the crushing elements through a material property definition. The input data for the crushing stress is usually based on the testing of flat test coupons, although the code is not tied to any particular testing method. The conventional failure mechanisms and damage models in all elements, both in and out of the crash front, remain active. If stresses in the element exceed the damage laws they can be degraded and failed part way through the crush process, thereby locally reducing the energy absorbed. In some instances it may be appropriate to use crush test data gathered from component tests if a number of similar component geometries have been crush tested to ascertain the characteristic crush stress. This is appropriate in situations such as braided tubes which cannot be tested as a sine or flat coupon.

As presented by Roberts [153] through the CMH-17 Crashworthiness Group, the crush stress is affected by the geometry of the crushing section. With the majority of crush materials, curved regions produce an elevated value of crush stress compared with flat regions. The elevated crush stresses associated with the curved or corner areas of a structure is thought to be due to the suppression of delamination by a hoop-type restraint and the addition of tearing-type failures. In the flat areas, the layers of the composite can delaminate more readily. The flat-region (or "free") crush stress can be determined by the testing of flat coupons crushed against a flat crush plate. The delamination-suppressed crush stress associated with curved or corner areas of a structure can be determined by crush testing of curved, self-supported crush specimens or by an adaptation of the flat coupon method that utilizes pin-stabilization in the crush plate to suppress delamination [73].

4.4.3 Delamination Modeling

The modeling methods described above all provide approaches for simulating the progression of a crush front in a composite structure. A related form of damage modeling involves the simulation of delamination growth occurring in a composite laminate as a result of the crush damage. If the laminate is modeled with multiple finite elements through the thickness, it is possible to consider the modeling of a delamination at element boundaries, and simulating the progression of delaminations growing along the element interface as a result of the crush event.

To model delamination formation and growth, interface elements, sometimes referred to as "cohesive" elements, have been developed and implemented into many commercial finite element codes. Although the use of interface elements does not require the modeling of an initial delamination, the possible delamination paths must be determined in advance and modeled with interface elements. Additionally, interface elements may only be used along existing finite element boundaries. The response of an interface element is defined by a traction-separation law as shown in Figure 4-21. This law assumes an initial linear elastic behavior followed by subsequent damage initiation and damage evolution. The elastic response of the interface element is defined by the user and established so that the initial compliance due to the interface elements is small compared to the overall compliance in the model. Damage initiation is determined by a user-prescribed critical stress P_c and represents the termination of the initial linear elastic response of the element. Damage evolution is defined as the region in which the stiffness of the interface element is reduced, corresponding to crack growth. This response is governed by the critical energy release rate, G_c.

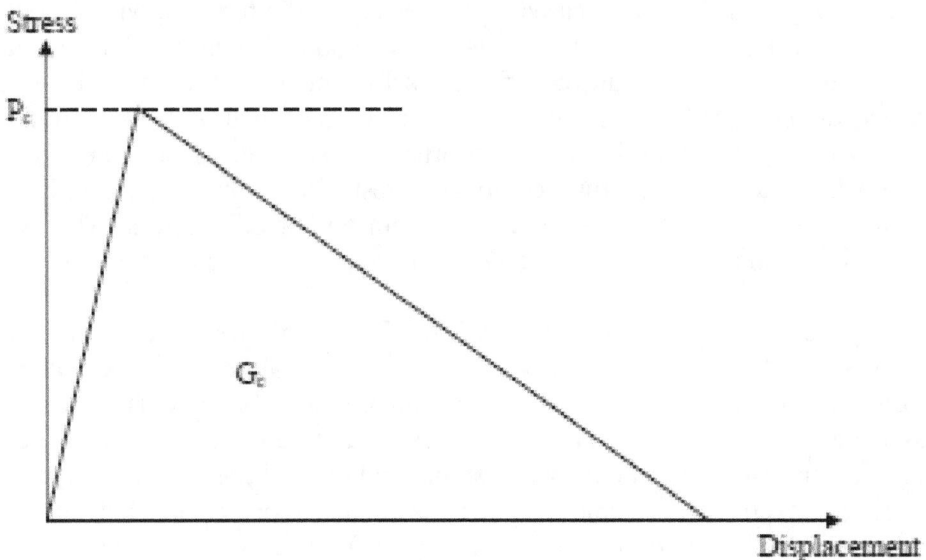

Figure 4-21. Traction-separation response of interface elements used for delamination modeling.

Typically the properties of the interface elements used to model delamination are determined by simulating the delamination propagation in a Double Cantilever Beam (DCB) specimen. A finite element model of the DCB specimen is created with this critical load that produced delamination growth is applied to the model. Analyses are performed with varying values of the critical stress P_c until crack propagation in the model occurs at the same applied load as in the experiment.

To date, limited crush modeling has been performed using interface elements to simulate delamination formation and growth during crush loading. Indermule [154] has used cohesive interface elements available with ABAQUS/Explicit in a crush simulation. Only limited results have been presented to date; however, the technique has been shown to be highly computationally expensive when used in a coupon-level simulation. As a result, it is unclear whether the use of interface elements to model delaminations associated with composite crushing are viable for automotive crash simulations.

4.4.4 Damage Modeling Away From the Crush Front

The concept of the crush zone and the need to maintain stability in the back-up structure (introduced earlier in this chapter) is an important consideration in crashworthiness modeling of composite structures. In order to effectively predict the crashworthiness of the structure, analysis techniques must be able to both model the progression of the crush front as well as ascertain whether the remaining structure behind the crush front will remain intact. In assessing the ability of a proposed modeling approach to predict failure of the back-up structure, it is essential that the forces entering the back-up structure from the crush are of a proper peak magnitude. As discussed previously and shown in Figure 4-19, the filtered force versus displacement results from crush modeling are not the internal forces that need to be internally reacted by the back-up structure during the simulation. While it is currently standard practice to compare filtered force versus displacement modeling predictions with experimental results to assess the accuracy of the crush zone modeling methodology, the unfiltered force versus displacement results must be compared to the experimental results when assessing the ability of a modeling approach to predict failure in the back-up structure. It is worth noting that the prediction of failure behind the crush front has been shown by Dodworth et al. [155] to be directly related to the ability to get the forces correctly into the structure.

To date, limited work has been published on the ability to predict failures of composite structures away from the crush front during explicit finite element modeling. Through the use of CZone, both MSC.Dytran and ABAQUS/Explicit codes have been used to determine the behavior of a complex composite crush cone during impact as shown in Figure 4-22a [83, 155]. Both codes take their forces from CZone in the crush zone and both utilize their native failure criterion and damage progression for the back-up structure. Both codes show a high degree of correlation with the experimental results for both forces and debris, the latter of which is shown in Figure 4-22b. Both the MSC.Dytran and the ABAQUS implementations predicted the pattern of debris produced in the actual test.

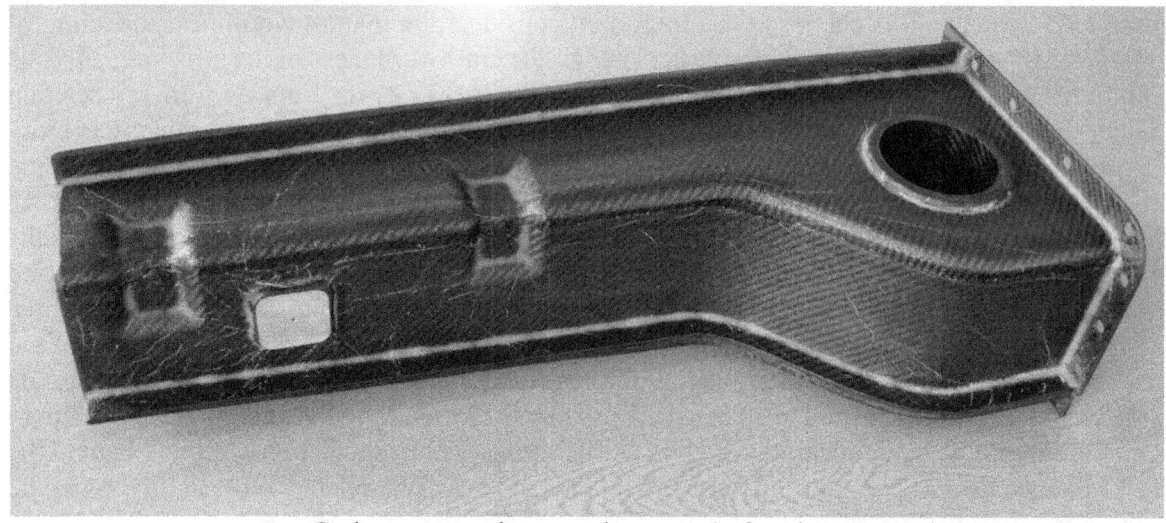

a. Carbon composite complex cone before impact testing.

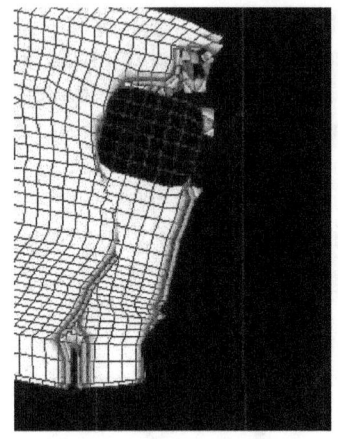 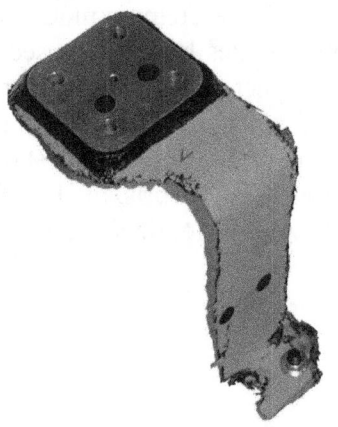 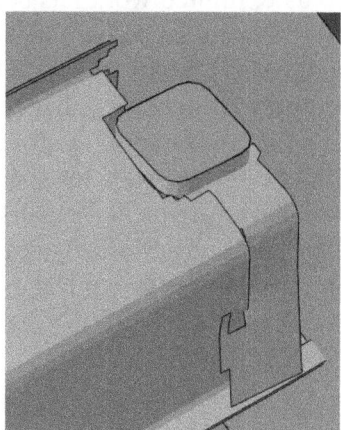

MSC.Dytran prediction   Experimental Result   ABAQUS prediction

b. Post impact condition.

Figure 4-22. Results of testing and analysis of carbon composite complex cone subjected to impact [83,155]

The MSC.Dytran failure approach used in the initial implementation was based on the Tsai-Hill [156] failure criterion, which covered all the conventional first-ply failure modes. But in addition, a degradation matrix is used, which links particular types of failures with the degradation of associated properties and, where appropriate, other corresponding ply attributes. For example, in matrix compression failure, the material constants E1 (longitudinal Young's modulus) are set to zero, but also the corresponding E2 (lateral Young's modulus), ν12

(Poisson's ratio), and in-plane shear are set to zero [157]. These properties are then degraded over a preset number of time steps.

One limitation of the MSC.Dytran implementation was the inability to handle internal material damping, but rather rely on system damping. The use of system damping suppresses artificial dynamic stresses internal to the shell elements, but artificially applies external damping to the elements which effectively leads to the whole structure moving through a highly viscous medium. However, ABAQUS/Explicit has a material damping feature that uses results from Dynamic Mechanical Analysis (DMA) testing. This feature allows for the required internal material damping to be applied to the model without the need for additional system damping.

In the latest implementation of CZone in Abaqus/Explicit, a different damage evolution approach is employed. The onset of failure in the back-up structure is determined according to a failure initiation criterion on a ply-by-ply basis. For woven composites with relatively brittle epoxy matrices, for example, the Tsai-Wu criterion [136] is used. However, the Hashin failure criterion [158] could be considered more appropriate for unidirectional plies within a multidirectional laminate. Once the onset of failure is reached, the properties of the element (and layer of that element) concerned are degraded according to a damage evolution model. This damage evolution approach causes a gradual (rather than instant) degradation of the ply stiffness, which better relates to the physical process of composite failure. The physical property of the damage evolution model is expressed in terms of an energy release rate in units of energy per area (e.g. J/m^2), which is obtained from mechanical testing. At the onset of ply failure (using the Tsai-Wu criterion [136] in the current implementation), the energy remaining is computed and a linear degradation of the ply level stiffness is applied. With continued load on the element (ply), degradation continues to the point where the measured energy at failure is achieved. If unloading occurs, the degradation is suspended and the element unloads and reloads from its instantaneous stiffness until it exceeds the degradation it is currently exhibiting at which point degradation resumes [159]. An implementation of this technique was also undertaken in LS-Dyna by Pinho et al. [134] and incorporated physically tested data for correlation based on Pinho et al. [160].

4.5 Other Failure Models

Although considerable amount of effort has been expended in developing failure criteria for composite degradation in the explicit regime, there is no agreed or accepted method being applied. Christensen [161] remarks:

> "Considering the difficulty of the topic, it is not surprising that there is nothing that is even within proximity of being a unified, verified, and reasonably recognized methodology."

Many codes and techniques have been specifically developed in order to predict the out-of-plane performance against penetration. Whatever the number of differing attempts at producing failure models that would be capable of determining first ply failure and subsequent degradation leading to full failure of the element, the authors are of the firm opinion that when utilized with the appropriate materials the failure models available in all the commercially available Explicit codes including LS-Dyna, RADIOSS, and PAMCRASH are capable of predicting failure in the back-up structure providing the correct loads are being predicted at the crush front, from the crushing simulation model.

The challenge is to utilize the most applicable failure theory for initiation for the materials in use or "engineer" and extend the domain of a failure theory to cover new and evolving materials to create a new failure theory. This is dependent on the software and the type of solution. On top of this the selection and method of degradation of progressive damage needs to be carefully considered [162].

4.6 Validation of Crashworthiness Modeling

4.6.1 Tube Testing

Perhaps the most common method for validating crash models for composites is through the simulation of crush testing of either square or round tubes of various fiber architectures. Two methods of tube crush testing may be performed: with or without a plug trigger. When performing the test without a plug trigger, the tube is impacted against a plate. Upon initiation of the crush front, the composite material comprising the tube would be subjected to the various crush mechanisms previously referred to in this document. Although this type of crush test generates the highest specific crush forces and hence the highest SEA, it has proved to be difficult to analyze, due to the apparent need to crush the element to zero size, as discussed previously. However, the Type 1 and Type 2 crush behavior exhibited in such testing is required to obtain the high energy absorption and light weight promised by the use of advanced composites. The alternative method of tube crush testing includes the use of a plug trigger. As described previously, the use of a plug trigger during testing produces axial tearing and the formation of fronds, which are forced around the radius of the plug and experience bending failure. Due to the different type of failure produced in the composite tube, the simulation of the plug triggered failure is a different challenge compared to the first case where the material is fragmented into small debris at the crush front. These types of failure are capable of being modeled with the failure models currently employed in the continuum and progressive damage models, with suitable tuning of the degradation variables. However, the use of a plug trigger generally does not produce Type 1 or Type 2 crushing, and therefore a tube test with a plug trigger is generally not a suitable test to be used for validate a modeling method to be used to simulate composite crush.

4.6.2 CMH-17 Crashworthiness Working Group Round Robin

The CMH-17 Crashworthiness Working Group is widely recognized as being a focal point for researchers focusing on advancing capabilities for composite material crashworthiness prediction. The main focus of the working group activities has been a numerical 'Round-Robin' activity, with participants invited to submit predictions of the crush behavior of several types of composite specimens. The working group meets every 6-9 months to review progress, and participants from a range of backgrounds and using several different software codes present and discuss their analysis predictions and their comparison with physical test results.

The CMH-17 crashworthiness working group launched the Round Robin process in 2007. In the first three years, progress has been hindered by the lack of crush-specific failure modes in the majority of the analysis codes. Various attempts have been made to modify and adapt existing composite failure models which have their origins in dealing with the onset of failure in composites. The work initiated in the CMH-17 Crashworthiness Working Group has initially focused exclusively on the crush front simulation and is scheduled to continue with this focus until December 2011 [163]. All the components tested to date have demonstrated a stable crush front, with no failures in the backup structure. As a result, these tests do not permit an assessment of a software's ability to predict crashworthiness, as the modifications made to the models to ensure stable crush are not being linked to the corresponding effect on the fidelity in predicting the strength of the back-up structure. Figure 4-23 shows two predicted failure progressions presented by Feraboli [164]. The first, shown in Figure 4-23a, illustrates a progressive crushing and is referred to by the author as a "Desirable Failure Mode." The second, shown in Figure 4-23b, illustrates an unstable collapse produced behind the crush front as is referred to as an "Undesirable Failure Mode". This change in performance in the model is due to the change of one "non-physical parameter" which is fundamental in controlling the energy absorbed, yet, digitally changes the failure mode.

In the pursuit of ever more efficient crash structures, engineers and material scientists are seeking materials and lay-ups with ever higher crush stresses, which in turn create proportionally higher stresses in the backup structure. These stresses, in conjunction with geometrical stress concentrations or buckling limits, can cause un-wanted failure behind the crush front. For a meaningful crashworthiness simulation it is essential that the analysis be capable of capturing both the crushing forces and the behavior, whether desirable or undesirable, of the backup structure without the need to tune parameters to achieve the desired response.

a. Progressive crushing.

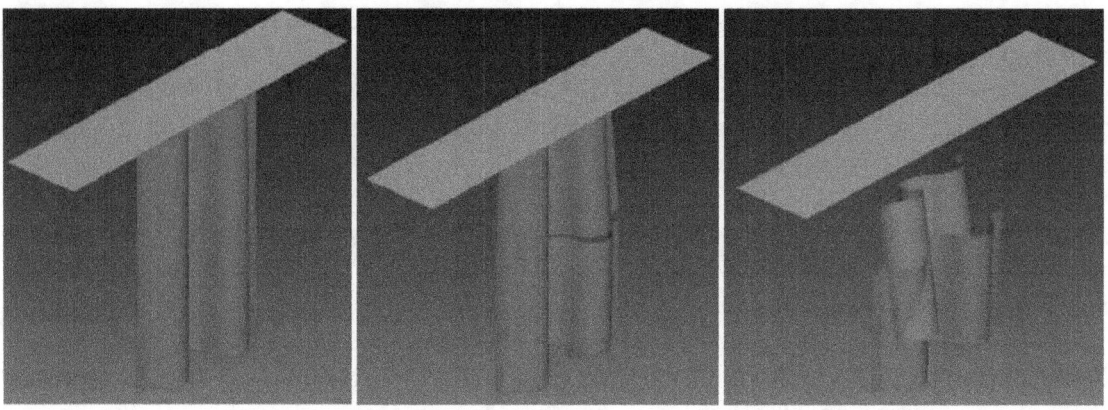

b. Unstable collapse produced behind the crush front.

Figure 4-23. Predicted failure progressions from coupon testing [164].

For the Round-Robin, all participants are provided the basic mechanical properties of the AGATE T700/ 2510 plain weave carbon/epoxy material along with optional flat plaques to undertake code specific material testing. Additionally, crush test results from sine wave specimens fabricated with a $[0/90]_{4s}$ 1.7 mm (0.065 in.) thick lay-up were supplied. It is noted that the 0/90 cross-ply lay-up is not typical of the lay-ups used in typical automotive crash members, which typically have off-axis layers.

The first phase of the Round Robin was to replicate the experimental crush force versus displacement results for the quasi-static crush test performed with a crush distance of approximately one half of the sine specimen's length. This initial phase was intended to allow material, contact, and simulation parameters to be tuned and set prior to analysis of more complex sections and scenarios. Although the testing was performed quasi-statically, all participants were required to process their solutions at 3 m/s (10 ft/s) in order to minimize analysis time. All participants were required to filter the output forces with a SAE600 filter; however, a number of participants have in addition published the unfiltered data.

The second phase of the Round Robin initially involved the prediction of five sub-sections of a square tube with the same $[0/90]_{4s}$ lay-up of the same material. After the conclusion of the March 2009 Crashworthiness Working Group meeting, participants were provided the experimental results of the second phase tests performed under quasi-static crush, as this was a necessary requirement for a number of the participants in order to tune their model and material parameters with the results of the changing sections.

4.6.2.1 Progress as of Atlanta CMH-17 Meeting, Nov 2009

The approaches to the solutions can be broken down into three broad categories; Shell Models, Multi-layer Shell Models with cohesive failure and a Multi-layer Solid Model. A brief summary of the approaches and the achievements is presented below.

Shell Approach

In principle the shell approach is highly attractive as it utilizes the model construction method used in all vehicle crash simulations conducted today. It is computationally efficient and the model construction technique is familiar to many thousands of experienced automotive crash engineers. Unfortunately, the simplicity of the shell formulation is a limiting factor in their ability to represent the physical mechanisms occurring in the crush front whether using continuum or progressive damage models regardless of their derivation either multi-physics or through testing.

The participants using the shell approach have predicted a 'saw-tooth' type force versus displacement response during crushing, with peaks well in excess of the physical test result and troughs well below. In cases where the input parameters are properly selected and the results filtered, the force versus displacement predictions are in general agreement with test results. However, the filtering masks the true nature of the forces acting on the structure. In the actual crush test being simulated, the force level maintains a more constant level with peaks that are on the order of 20% to 30% above and below the average. Many of the simulations in their true unfiltered form give peak force levels which 'saw-tooth' from 0% to 200% of the average force level. This pattern of oscillating forces is governed by the finite element mesh, with each row of crushed elements giving rise to a new peak before a failure condition is reached and the element is deleted from the analysis. The only participant able to overcome this issue is the Abaqus+CZone submission presented by Roberts [165]. In this simulation the crush forces are calculated and applied using a crush-zone contact model based on the projected crush area of the shell elements at the crushing interface. The model inputs are obtained from the progressive crush data from both the flat and the delamination-suppressed sine testing undertaken by Roberts and Barnes [73] and Feraboli [166]. The crush forces from this simulation remained at a constant level established using input data, and no unrealistically excessive forces were evident in the unfiltered data. The CZone solution also maintains the damage evolution models in all elements in order to be able to simulate additional non crush failures that evolve. The ABAQUS+CZone simulations of five Round II shaped specimens were shown to be performed in a similar manner by applying the crush data to the corners and flat portions of the shaped specimens in a consistent manner. These solutions were also shown to be consistent with changing mesh densities and ill-conditioned meshes [165].

Multi-Layer Shells

The PamCRASH submission [167] of the Round I sinewave specimen was presented for the first time at the Atlanta meeting. Previously, simulations had been performed on a specific DLR specimen. The technique differs from that used by other participants as it requires the user to pre-identify the damage mechanism and damage extent in the mesh. Effectively, a "fragmentation wedge" is modeled between the impactor and the structure. The resulting force levels were highly oscillatory, showing a similar 'saw-tooth' characteristic discussed above. The need to pre-specify the crush wedge at the damage zone prevents the simulation from predicting the performance of any structure that has a non-crush failure behind the crush front, which is evident in many real impact situations.

The ABAQUS Continuum Shells submission [154] was briefly previewed at the Salt Lake City meeting for the first time. Each ply of the laminate is modeled as a separate layer of shells. The elements are represented as solids with eight nodes, but are indeed computed using standard shell theory for economy. Each element is connected to the adjacent ply by a cohesive interface. Crack propagation at the damaged cohesive interface is controlled by an interface fracture energy criterion. Results from the simulation were incomplete, and conclusive comments on its capability are therefore difficult. It is understood from the presentation, however, that the model has suffered from stability issues and long run times (several days) and that an analysis has yet to run to completion.

Multi-Layer Solid

Only one participant utilized solids, LS-Dyna MAT162 [168] using a specialist material model developed for progressive failure in composite materials subjected to high strain rate and high pressure loading. The model is meshed with solid elements without an interface element on a one element per ply basis and a 2.5 mm (0.10 in.) element length. Despite its high computational overhead and modeling complexity, this model has is believed to have the potential for capturing the physical effects of the crush interface. This conclusion is based primarily on the review of its out-of-plane loading performance and not the specific axial crush examples of the CMH-17 Round Robin. However, there are several unmeasured model calibration parameters that are deployed. For example, "sffc" appears to be a non-physical parameter which is only active on the compressive stress property. Only the C-channel specimen from the Round Robin II was modeled, and the model is still referencing non-physical parameters after the Round Robin I exercise in order to achieve improved correlation. According to the author, *"Significant work still required on Round Robin case studies to improve correlation and establish degree of sensitivity of MAT162 model parameter* [168].*"*

4.7 Summary: Composite Crashworthiness Modeling

The CMH-17 Crashworthiness Round Robin has been an important focal point in bringing together researchers attempting on advance capabilities for crashworthiness of composite materials. However, the small-scale nature of the current Round Robin specimens being simulated is still a long way from the reality of vehicle crashworthiness prediction. To predict vehicle crashworthiness it is necessary for the numerical technique to predict how a component, sub-system and vehicle structure will behave in a crash. Depending on the design and material characteristics the real-life behavior may involve the desired progressive crushing mode, but also other undesirable failures such as buckling or compressive failure; there may be transitions from one kind of failure model to another as the geometry, load paths and component interactions determine during the crash event. The successful simulation technique for vehicle crashworthiness must handle all of these uncertainties and eventualities within practical simulation timescales. Specific recommendations regarding the development of modeling methods for composite crashworthiness are presented in Chapter 5.

5. RESEARCH NEEDS FOR MATERIAL DATABASES, TEST METHOD DEVELOPMENT, AND CRASH MODELING

5.1 Introduction

As discussed in Chapter 4, progress continues to be made in the areas of materials databases for composites, test method development for composite crashworthiness, and crash modeling of composite structures. However, significant needs exist in all three areas. In this chapter, the current status of these three topic areas pertinent to crashworthiness is summarized. Recommended research and development efforts needed in each of these three are provided.

5.2 Material Databases

5.2.1 Current Status

As summarized in Chapter 4, the majority of the progress to date in developing databases for composite materials has been associated with standard stiffness and strength, or "mechanical" properties of composite materials. Additionally, the relatively small number of composite materials with commercially-available, well-populated material databases are considered "aerospace-grade" composites, composed of high performance and relatively expensive fibers and resins, manufactured using aerospace-type manufacturing methods. Such material systems are generally not well suited for cost-conscience, high-volume automotive applications.

Of the high performance composite materials with relatively extensive databases, a majority are either company proprietary or have restricted distributions such that they are not publicly available. However, shared databases for several high-performance composite material systems are available through the Advanced General Aviation Transport Experiments (AGATE) database. As described in Chapter 4, these composite materials were selected based on the needs of the general aviation community. Additionally, several high-performance composite material systems are currently being characterized through the National Center for Advanced Materials Performance (NCAMP), located at the National Institute for Aviation Research at Wichita State University. These composite materials that are currently characterized as a result of both AGATE and NCAMP activities utilize high cost materials and processes, and are considered to be of limited usage for automotive applications.

Among the lower cost composite materials that are of greater interest for use in automotive applications, limited material property data is currently available. In general, limited data is available for most commercially available composite materials from the material supplier. While the representative mechanical properties provided may be suitable for material selection and possibly initial design, the available data are rarely sufficient for design and static structural analysis purposes, and insufficient for dynamic crush analyses. Therefore limited mechanical property data is currently available for composite materials of interest for automotive applications.

Of the currently available material property databases for composite materials described above, none include the specialized crashworthiness properties needed for automotive applications. The two primary crashworthiness properties of interest, discussed previously in Chapter 4, are the Specific Energy Absorption (SEA) and the sustained crush stress. These properties require additional crush tests that are currently not part of any standard composite material characterization programs, and are typically not available through material suppliers. Two additional properties of interest for some composite crash modeling approaches are a measure of damping and the in-plane fracture toughness of the composite laminate (associated with tearing-type fracture). Neither of these additional properties are commonly available in current material databases. However, DMA testing, commonly used to establish the glass transition temperature (T_g) when characterizing composite materials, can also be used to provide a measure of material damping with a suitable test procedure.

While a number of organizations and corporations have compiled crashworthiness data for composites, they are not generally available or in the public domain. One example of a known material database that includes specialized crashworthiness properties of composites is that which is under development by the Automotive Composites Consortium (ACC). This database contains material data generated through ACC research activities. Since such research activities often utilize specialized tests, the data generated is generated is generally not intended for the general characterization of crashworthiness. As such this database is perhaps better classified as a data "depository" rather than a database of standard types of data. As mentioned previously in Chapter 4, the development of this database is nearing completion, and will be available to ACC participants but not to the general public.

5.2.2 Recommendations

As composite materials continue to be identified for potential use in automotive structures, specialized testing will be required to measure crashworthiness properties required for both qualitative assessment of crashworthiness as well as quantitative properties required for modeling. Recommendations regarding testing and modeling will be discussed in subsequent sections. However, specific recommendations concerning the development of material databases for composite crashworthiness are discussed below.

5.2.2.1 Identification of Required Properties for Crashworthiness Databases

The first recommendation is that a consensus opinion be developed on which properties are required for inclusion in a material database for use in automotive crashworthiness. As a starting point, those properties required for automotive design and analysis using conventional static and fatigue analysis should be considered. Additionally, those properties included in composite material databases for aerospace-grade composites should be considered. However, not all of the properties included in these two sources may be needed, as they may be specific to the crashworthiness of metals or requirements specific to aerospace applications, respectively. Added to this listing of properties should be placed the specialized crashworthiness properties required for composites, which should be obtained at elevated impact velocities. The following specialized crashworthiness properties are recommended for inclusion:

1. Sustained Crush Stress: This property has been used in the preliminary design and development of crush structures for motorsport and automotive applications for many years. Usages include both hand calculations and as input into finite element linear static simulations of the back-up structure. The Sustained Crush Stress property is also used directly in at least one modeling approach (Phenomenological approach, ABAQUS+CZone), where this measure of crush stress is considered a laminate property, similar in importance to the compressive stress.

2. Specific Energy Absorption (SEA): Although the SEA is currently the most recognized measure of the crashworthiness of a composite material or laminate, its usefulness typically is limited to material and laminate screening and ranking purposes.

3. Compression Crush Ratio (CCR): The CCR is defined the ratio of the compressive strength to sustained crush stress of a composite laminate. Used as an indicator of the likelihood of crushing in a stable manner, the CCR is the primary property used for predicting failure of the composite component behind the crush front.

4. Measure of Crush Stress Variability: In addition to the sustained crush stress, a measure of the variability in the crush stress during crushing is recommended. While some materials exhibit a relatively constant stress level during crushing, others exhibit erratic behavior. This variability results in peak stresses well above the average value reported as the sustained crush stress. A variability measure is believed to be of use for assessing the likelihood of failures behind the crush front; the higher the variability, the higher the peak stresses and the greater the likelihood of a failure occurring in the back-up structure.

5. Force Versus Displacement Plots: Since the needs of different crashworthiness modeling approaches vary considerably, it is recommended that the actual force versus displacement test data (in unfiltered form) be included in the material database, such that other modeling-specific properties may be obtained or required model parameters be calibrated.

5.2.2.2 Crashworthiness Screening Testing of Candidate Composite Materials

The second recommendation is that a program be initiated in which appropriate screening tests be performed using composite material systems that are viewed as viable for PCIVs. Such composites would include affordable material forms and manufacturing processes.

Since crashworthiness of a composite laminate is a function of the thickness and stacking sequence of plies, screening tests may need to be performed on a variety of laminates, suitable for use in different components. While such a characterization may seem to be large in scope, relatively simple coupon-level testing may be used for such screening tests. As a starting point, it is recommended that the SEA and the sustained crush stress be measured for a variety of materials (including carbon and glass fibers as well as multiple resin systems), multiple fiber forms (including continuous and chopped fibers), and textile preforms (including weaves and braids). Such testing would need to be performed for specified fiber orientations, specimen thicknesses, and fiber volume fractions that are identified as viable for specific automotive components. Measurement of the compression strength of the material is desirable such that the Compression Crush Ratio (CCR) may be calculated. The CCR is an important factor in determining the likelihood of the composite material to behave favorably in an actual automotive application.

An additional complication when screening composite materials for crashworthiness is the issue of crush velocity effects. Several researchers have observed significant crush velocity effects on sustained crush stress measurements during crush testing [93, 95, 110, 116, 119]. Typically, a relatively constant sustained crush stress is observed at velocities above a transition point which typically occurs in the region of 0.5 to 1.0 m/s (1.6 to 3.3 ft/s). Below the transition point, elevated crush stresses are observed. It is not uncommon to see amplifications of twice the dynamic sustained crush stress. As a result, quasi-static testing is widely viewed as inappropriate for assessing the crashworthiness of candidate composite material systems. Additionally, minimal published data exists that may be used for material screening at higher crush velocities and therefore higher material strain rates. It is recommended that screening testing be performed at multiple crush speeds. Attention should be given to the actual crush speed during testing in addition to the initial impactor velocity, since the impacting mass decelerates during drop-weight impacting and a portion of the crushing may occur both above and below the transition described above. Both the SEA and the crush stress should be measured in the velocity regime close to the peak velocity of the impact, where the majority of the impact energy resides.

5.2.2.3 Development of Material Database for Crashworthiness Model Development

The third recommendation is that a complete material database be developed for one or more composite material systems deemed as well-suited for usage in future PCIVs. The initial purpose of this database development effort would be to validate crashworthiness model development efforts as well as to tune any required model parameters. The choice of the composite material(s) should be based on previous experiences within the automotive industry as well as screening testing performed on a variety of composite materials. Ideally, the database would include all of the properties identified in Recommendation 1. Additional specialized crashworthiness testing may be required to assist in model development, including additional coupon-level and element-level testing. When possible, all crashworthiness testing should be performed using the same laminate and specimen thickness, as described in Recommendation 2.

Depending on the modeling approach to be used, the needs for specialized material data can vary tremendously. For example, ABAQUS+CZone requires only the sustained crush stress and the material damping. Other approaches currently require a complete force versus displacement response from a similar structure, such that crush parameters in the model can be calibrated. In addition, all modeling approaches require material strength data to compute the failures in the back-up structure (away from the crush front).

When developing material databases, it is important that testing be performed using the same test methods, procedures, and data recording and reduction methods. In the absence of standardized test methods, it is preferable that such material databases be developed by a small number of coordinated laboratories. When possible, data should be developed using standardized test methods and procedures, and reduced using standard practices developed in accordance with CMH-17 data processing procedures. Currently, however, there are no test methods that have been universally accepted or standardized for obtaining crashworthiness properties. Current status and future recommendations on test method development are discussed in the following section.

5.3 Crashworthiness Test Methods

5.3.1 Current Status

As summarized in Chapter 4, significant progress has been made in composite crashworthiness test method development at both the coupon-level and element-level. However, no standards currently exist by which either type of testing may be performed. As described in Chapter 4, "coupon-level" testing utilizes a relatively small test coupon and is intended to characterize the crashworthiness properties of a composite laminate - independent of any structural-level features that may be found in an intended application. In contrast, element level testing utilizes a test article that is intended to be "representative" of the intended application. The current status of both types of testing are summarized in the following sections

5.3.1.1 Coupon Test Methods

Within the category of coupon-level test methods, development efforts are currently continuing in both the self-supporting coupon and flat coupons categories. While self-supporting coupons do not require a specialized test fixture to achieve stable crushing, specialized fabrication is required to produce the required out-of-plane curvature in the shaped specimens. Additionally, the effect of the specimen shape on the test results remains a concern, as it is generally not possible to produce the same geometry as is present in the intended application, within the coupon. As summarized in Chapter 4, several self-supporting coupon geometries have been proposed for crashworthiness studies. No standardized specimen shape exists; in fact some of the geometries have been selected with no intention of standardization. Of those intended for general investigation, the sinusoidal-shaped specimen appears to currently have the greatest level of interest. Although the test results are not directly applicable to a structural application, they can be viewed as representing the laminate property for the delamination-suppressed or curved regions of a structure. However, such specimens do not provide results pertinent to the flat regions of a structure. Furthermore, the nature of the geometry makes it difficult to change specimen thickness, as the surfaces angled to the mold draw direction have a different relative thickness increase to those parallel, leading to a variable volume fraction through the component. However, self-supporting coupon tests are simple to perform quasi-static tests.

In contrast, flat coupons do not require specialized fabrication; specimens may be machined from conventional flat plaques. However, a specialized test fixture is required to support the specimen during crush testing. The use of flat coupons and specialized test fixtures is consistent with other characterization testing for composite materials, as virtually all of the properties listed in AGATE and NCAMP databases as well as material datasheets from suppliers are obtained using flat coupons.

While flat coupons are not representative of all areas of the geometry found in automotive structural components, they are useful for studying laminate crush characteristics, making relative comparisons of composite materials and fiber architectures, and obtaining input data for computational models. Additionally, results from flat coupon testing may be used to predict aspects of structural behavior, as structural components typically have regions of flat geometry as well as regions with curvature such as corners. Results from flat-coupon crush testing are directly applicable to flat regions of a structure. Similarly, results from self-supporting specimens may be applicable to the curved regions of a structure. The primary differences between the response of these two different regions is believed to be the degree to which delamination is allowed to propagate and the amount of tearing produced during crushing. Curved specimens tend to produce higher values of SEA due to the suppression of delamination as well as tearing at locations of curvature change. In contrast, flat coupons do not produce delamination suppression and do not experience tearing, yielding lower values of SEA.

As described in Chapter 4, several flat-coupon test fixtures have been proposed in recent years, some of which are currently in use. The specialized test fixtures currently being considered all incorporate an unsupported region for specimen crushing. In an attempt to simulate the higher SEA associated with a curved portion of a structure while using a flat coupon, Roberts and Barnes [73] are currently developing a "pin supported" flat coupon test, which utilizes the same test fixture but a different support base. The pin-supported base requires the specimen to tear around the pins, similar to in the sinusoidal specimen, and also serves to suppress delamination. As a result, a flat coupon may be used to simulate the crush behavior of both the flat portions and curved portions of a composite structure. This pin-supported flat coupon test method is currently under development and is producing comparable sustained crush stress results to the sinusoidal and tube specimens [73].

5.3.1.2 Element-Level Test Methods

Of the element level test method investigated to date, untapered tubes of either square or circular cross section are most commonly used. As presented in Chapter 4, research on the development of tube crush testing has been ongoing since the late 1970's and continues to be a common practice for characterizing energy absorption for automotive applications. Of the two cross sectional shapes, circular tubes have been found to produce the largest value of SEA. Square tubes, however are often more representative of automotive structural frame components. To date, no standardized test methods have been developed for composite tube crush testing.

In general, no specialized test fixturing is required since tubes are self-supporting structures. In some cases an external plug trigger is used to promote progressive crushing. The use of plug triggers with circular or square tubes typically results in axial tearing and formation of fronds, which undergo a continuous bending failure as they are driven onto the radius of the plug. As a result, tube crush testing using plug initiators typically produce lower values of SEA than flat-coupon tests or a flat impactor on the same tubes.

To assess the ability of candidate modeling approaches to predict failure in the back-up structure during a crush event, additional specialized testing will be required. Such testing could include the use of tubes with tapered thickness or section increase with one or more stress concentrations (holes) behind the crush front. The test articles could be designed to transition from crushing at the crush-front to failure in the back-up structure either with increasing crush distance or using increasing hole sizes.

5.3.2 Recommendations

Although significant progress has been made in recent years towards the development of crashworthiness test methods, no standardized test method currently exist. Specific recommendations regarding the development of standardized tests for composite crashworthiness are discussed below.

5.3.2.1 Further Development and Standardization of a Flat-Coupon Composite Crashworthiness Test Method

The first recommendation is that a flat-coupon test method be developed and standardized for use in assessing the crashworthiness of composite materials. Based on best-practices established to date, the flat coupon test method should incorporate the following considerations:

- Means of accommodating different coupon thicknesses
- An adjustable, unsupported gap region in which crushing can take place
- No knife edge supports which introduce tearing
- Ability to be used quasi-statically as well as dynamically on drop towers and hydraulic test frames
- Ability to include different striker plates for pin stabilization or surface finish

- Easy to use
- Optional ability to view the crush zone of coupon during testing for high speed video

As part of development of a standardized test method, it is anticipated that a method of normalizing the test data will be required to account for the vibrations produced in the test fixture during dynamic loading (rig ringing). One goal of such normalization would be to produce the same force versus displacement response when the test is performed in different drop towers or servo-hydraulic impactors.

Additionally, it is recommended that the pin-supported base concept, recently introduced by Roberts and Barnes [73], be incorporated into the flat-coupon test fixture and further developed for use in simulating the crush failure modes exhibited in regions of curvature in structural applications.

Finally, it is recommended that the flat-coupon test methodology be adaptable such that an untabbed compression test can be performed using the same fixture and general test methodology, utilizing a tapered-width compression specimen. This compression testing, which may be performed quasi-statically using the same test panels from which the crush specimens are machined may, be used for determining the Compression Crush Ratio.

5.3.2.2 Further Development and Standardization of a Tube Test Method

The second recommendation is that a tube test method be developed for use in assessing the crashworthiness of composite materials. Although considerable composite crashworthiness research has been performed to date using tube testing, no consensus currently exists regarding several aspects of tube test methods, including the most appropriate crush trigger, the best method for force measurement, and proper data reduction and filtering methods. It is recommended that further research be performed to address these outstanding issues, develop a consensus opinion on best practices, and draft a standard test method.

Subsequent to this tube test development effort, it is recommended that a modification to the tube test be developed for use in assessing the ability of candidate modeling approaches to predict failure in the back-up structure during a crush event. It is suggested that the tapering of a tube, either the diameter or the thickness, be utilized to initialize crushing behavior at the crush front, followed by failure of the tube at a stress concentration (such as a hole or a series of holes) located some distance from the crush front. An increasingly higher force level will be required to continue crushing, leading to the eventual transition to a failure of the back-up structure.

5.3.2.3 Testing for Material Screening and Crashworthiness Model Development

As discussed in the Material Database section, testing is recommended using the proposed standardized test methods for purposes of both material screening as well as to support model development efforts. Initially, it is recommended that such testing be performed by a small number of coordinated laboratories, such that any potential problems can be identified, addressed, and rectified in the development of the standardized tests. As discussed previously, additional specialized tests may be required to support the development of specific modeling approaches. For example, Roberts and Barnes [73] have developed a simple tapered cone test article with both circular and rectangular cut-outs to produce buckling and other out-of-plane failure modes.

As initial testing progresses using the proposed standard test methods, industry standard Non-Destructive Inspection (NDI) should be performed on test panels prior to coupon cutting in an attempt to begin to address the effects of laminate quality (particularly the presence of voids) on crush performance of corresponding individual coupons. Such determinations will help to build a significant database for quality manufacturing purposes when deployed in full-scale production.

5.4 Crashworthiness Modeling

5.4.1 Current Status

Crash modeling of composite structures continues to be an important research area towards the development of PCIVs. Since composite crashworthiness modeling efforts generally utilize existing explicit finite element codes that have been used previously for crash modeling of metallic components, the primary focus of current modeling development efforts involves modeling of the crushing phenomenon occurring at the crush front in an experiment. As discussed previously in Chapter 4, the crush phenomenon in composites is widely accepted as involving different failure modes than those observed in conventional metallic materials. Although it has received considerably less attention, modeling the initiation and progression of damage away from the crush front in the back-up structures also remains an important consideration. While both capabilities are absolutely essential for crashworthiness modeling of composite structures, modeling of the crush front behavior has received the majority of the attention to date, since virtually all of the experiments used for validation of crush models have included only progressive crush at the crush front. For a meaningful crashworthiness simulation, however, the behavior of both the crush front and the back-up structure must be predicted without the need to tune parameters to achieve the desired response.

As described in Chapter 4, several modeling approaches are currently under development for crashworthiness modeling of composite structures. A majority of these modeling approaches are currently being evaluated through a numerical "Round Robin" activity as part of the CMH-17 Crashworthiness Working Group. Round Robin participants are invited to present and discuss their analysis predictions and comparisons with physical test results. The round robin was initiated in 2007, and to date has focused exclusively on the crush front simulation [163]. Participants from a range of backgrounds and using several different software codes and modeling methodologies are currently participating in the Round Robin, making this activity a useful tool in assessing the current status of crush modeling development efforts for composites. The "current status" of these modeling efforts is based primarily from the full-day

Crashworthiness Forum that was part of the November 2009 CMH-17 meeting [169]. However, the current status of other crashworthiness modeling efforts known to the authors that are not represented in this Round Robin activity are also summarized.

In the first phase of the numerical Round Robin, participants were requested to replicate the experimental crush force versus displacement results obtained from quasi-static crush testing of a sinusoidal coupon. Participants were provided with mechanical properties of the plain weave carbon/epoxy material along with optional flat panels to perform any code-specific material testing. The force versus displacement results from testing were provided. As a result, this initial phase of the numerical round robin allowed all participants to calibrate or tailor their models such that their numerically-produced force versus displacement results would be in agreement with that obtained from testing.

As expected, all participants simulating the Phase I test were able to obtain force versus displacement results that were in reasonable agreement with experimental results. A majority of results presented were filtered (discussed in Chapter 4), eliminating excessive peaks and valleys in the force response. In fact, a filtering frequency (600 Hz) was prescribed for Phase I analyses. While not determining whether any of the modeling approaches could be considered as predictive (since the response was given and allowed to be used in developing the prediction), this initial exercise did allow participants to establish suitable values of any required modeling parameters as well as gain experience applying their modeling approaches to the crush characteristics of the woven carbon/epoxy composite material used in the experiments.

Round II of the numerical Round Robin was originally intended to establish whether the modeling methodologies being used by the participants exhibited predictive capabilities for crush front modeling. To do so, the modeling method would be required to predict the force versus displacement response when experimental data was not provided in advance. Participants were requested to submit simulation results corresponding to five sub-sections of a square tube with the same layup and the same material used in the Phase I sinusoidal coupon test. Initially, participants were not given experimental results. After the conclusion of the March 2009 Crashworthiness Working Group meeting, however, participants were provided the experimental results of the Phase II experiments so that participants could assess the degree of correlation with their model and, if necessary, further tune their model and material parameters using the results of the five different sub-sections. The following summaries of the Round II modeling activities, based on presentations made at the November 2009 CMH-17 Crashworthiness Working Group meeting, are believed to represent the current status of these modeling approaches.

Participants using LS-Dyna with either the MAT54 material definition [170] or MAT58 [142, 171] showed that it is currently not possible to obtain good correlation with experiment for different structural shapes with constant values for material properties and crush parameters. Thus, the use of these two modeling approaches cannot currently be considered a predictive capability. Xiao [142] showed that using the constant value of the SOFT parameter from Round I correlation resulted in poor correlation between predicted and measured force versus displacement response. Deleo et al. [170] confirmed this result for the MAT54 material definition, showing that different values of the SOFT parameter must be used for each of the structural geometries. A possible correlation between the cross-sectional shape and the soft parameter is currently being explored. Currently, however, there is insufficient information to suggest that the use of either of these material definitions with LS-Dyna can be considered a predictive tool for the force versus displacement response corresponding to composite crush.

Force versus displacement results presented from the RADIOSS Ford modeling approach [172] were in general agreement with experimental results. The Material Law 25 failure law deployed fails a ply when either the 1 or 2 direction tensile stain allowable is exceed, and the physically predominating compressive strains do not initiate a brittle failure response. As with other continuum damage modeling approaches, the force versus displacement results presented were filtered. As discussed in Chapter 4, these modeling approaches are afflicted by oscillations in the crush forces significantly above the experimental results. With the aid of filtering, these extensive oscillations are reduced to levels that are in general agreement with experimental results. However, such peak forces would be destructive unless failure in the back-up structure is not suppressed. Without a knowledge of unfiltered results, it is difficult to assess the viability of such methodologies presented with respect to predictive crashworthiness capabilities.

For the RADIOSS Altair modeling approach [173], results were presented only for the Round I sinusoidal specimen and the Round II C-Channel section. To achieve a stable solution when modeling the C-Channel section, the author decreased the contact stiffness by a factor of 10 and increased the wall mass by 500 kg (1,100 lb). The developments of the solutions were dominated by a tensile failure mode (despite a compressive loading regime). To increase the energy absorbed and reduce the initial peak load, an investigation was undertaken in which a 0.01 mm (4×10^{-4} in.) random perturbation was introduced into the regular 1 mm (0.04 in.) mesh size. This perturbation in element side length caused catastrophic instability. The mesh size of 1 mm (0.04 in.) and resulting solution time step produced considerable computational expense that is in excess of current demands of the conventional solutions.

The ABAQUS+CZone modeling approach [174] was used to model all five of the Round II specimens. Crush stresses, obtained from testing of both flat coupons and the available sinusoidal specimen crush data, were applied to the corners and flat portions of the structures in a consistent manner. With a 5 mm (0.20 in.) nominal mesh size, the solutions were obtained for all five specimens concurrently in 46 minutes. Since the material crush characteristics were input to the analysis based on an average crush stress, the models predicted a noiseless (level) crush force entering the structure. Although actual crush testing exhibits a degree of oscillation in crush force, the average response of many coupons yields a somewhat level characteristic. However, the authors noted that if the variability in crush stress can be acquired and defined, the modeling approach can be extended to include the measured variability of crush stress.

For the PamCRASH modeling approach [167], results were presented for only the Round I sinusoidal specimen. The main failure mode shown for crush was dominated by delamination between the individually-modeled shell layers representing the plies. However, the geometry and location of the delamination "wedge" that is present in some crush observations was required to be pre-determined. The computational demands of his approach, which include interlaminar contact and multi-layer stacked shells, are also believed to be in excess of current demands of the conventional solutions.

Both the ABAQUS Continuum Shells submission [154] and the MSC MAT162 implementation within LS-Dyna [168] were categorized by authors as research tools, and neither were used to simulate the Round II specimens. As such, it is difficult to assess their current capabilities. However, both modeling approaches are believed to require significant computational resources.

5.4.2 Recommendations

Although considerable progress is being made towards the development of modeling approaches for predicting the crush behavior of composite structures, a series of recommendations are made for future modeling efforts as discussed below.

5.4.2.1 Further Assessment of Modeling Approaches for Crashworthiness Modeling

The first recommendation is that continued and expanded assessment be performed to assess the capabilities of current modeling approaches to predict the crush behavior of a variety of composite materials and structures. In coordination with the recommended testing described in the previous section, focus should be placed creating a suite of benchmark model validations supported with detailed experiments. These model validations would feature the use of simple models and would focus on assessing the predictive capabilities of several modeling aspects, including:

- Correctly predicting the peak force as well as the force versus displacement response due to Type 1 and Type 2 crush behavior, as presented in Chapter 4:

    Type 1: Fiber and matrix fragmentation characterized by small debris.

    Type 2: Significant delamination ahead of the impactor in flat coupons, formation of fronds in tubes

- Ability to integrate Type 3 failures, where the failure mode is essentially not crush, and significant energy absorption occurs due to bending failure away from the crush front.

- Reasonable run time – Comparable with metallic vehicle analysis resource requirements.

- Ability to have a single material property specification derived from testing regardless of evolving impact requirement and development of new crush fronts.

- Ability to accommodate different interfaces and crush initiators

Ability to replicate observed material damping for a variety of impact and crush scenarios from measured quantities or from an initial validation experiment.

5.4.2.2 Assessment of Modeling Capabilities to Predict Response of the Back-Up Structure During the Crush Event

The second recommendation is that additional assessment be performed to assess the capabilities of current modeling approaches to predict the response of the back-up structure in a composite automotive component during a crush event. The response of the composite structure away from the crash front has not been given adequate attention to date, and yet remains an important requirement for crashworthiness modeling and should not be disassociated from other crashworthiness requirements. Of particular interest is assessing the ability of candidate modeling approaches to predict the response of the back-up structure to several possible failure scenarios through forces generated within the crush front. Similar to Recommendation 1, these model assessments would be performed in conjunction with specialized experiments. As discussed in the previous section, such experiments may include a tapered tube with a stress concentration (hole) in the back-up structure.

Additional model validations would focus on assessing the predictive capabilities of several failures in the back-up structure, including:

- Predicting interlaminar failure – model delaminations forming and propagating behind the crush front.

- Predicting fastener failure.

- Address section stability – buckling of the test article.

- Predict fracture at an adhesive bondline in the structure.

- Address contact interaction with other structural members, particularly metallic components.

- Address pull-out or pull-through of inserts and onserts.

- Address low-cost rivets and other peel stoppers.

5.4.2.3 Accounting for the Stochastic Nature of Crush Force Inputs and the Factored Allowables in the Back-up Structure

In order for automakers to gain confidence in the application of composites in automotive structures, it is necessary to demonstrate reliability and as well as the ability to predict the reliability for a production environment. Within the aerospace industry, considerable emphasis is placed on the confidence of static material properties of composite materials using A-basis and B-basis allowables. Unlike most analyses where the load input can be represented by a maximum value, however, crushing is effectively a load-limiting case, and therefore cannot simply be factored to provide confidence. The inherent variability of the crush load needs to be accounted for in the simulations, as this has the corresponding effect of the fidelity on the prediction in the back-up structure.

An extension to the assessment of the modeling approaches is recommended that includes the capability to adequately represent the fluctuation in crush forces entering the back-up structure. This recommendation should be linked to Recommendation 1 to allow a demonstration of the capability on a tapered tube with a hole, where a number of samples can be tested and the statistical confidence in the crush force input and the material allowable can be evaluated.

5.4.2.4 Development of a Reusable/Universal Benchmark System(s) for Crashworthiness

It is clear that the maturity of software and materials testing approaches for composite impact analysis significantly lags the demand from the automakers. As a result, many different avenues have been pursued by different organizations in an attempt to validate their individual approaches to simulation. With a number of test methodologies being developed and the commercial availability of software specifically designed for the purpose of dynamic impact, it is recommended that a comprehensive and independent benchmark problem be developed for the assessment and validation of the steps required to develop future PCIV body structures.

To overcome the difficulty presented by the differences in the currently evolving methodologies and techniques, it is recommended that a universal benchmark be established that can be used for the testing, the software, and even the analyst. The recommended benchmark should be on an automotive scale and address automotive requirements. Most importantly, it must have the ability to be openly solved and discussed, without prejudicing the provenance of the exercise for subsequent participants. Unlike conventional prescribed benchmarks, this approach encourages development of material, testing, and analytical solutions which can be openly assessed as it allows various parties to offer different solutions to the same fundamental problem.

The recommended benchmark problem consists of a structure designed to absorb significant energy in an impact. A likely structural component to select would be the front longitudinal of a vehicle, in a two-part bonded assembly. The external tool geometry will be fixed and this will be owned and controlled by the custodians of the benchmark. Along with the exterior geometry, a clear performance specification will be provided for the energy absorption and static performance of the structure, which will be comparable to a PCIV longitudinal requirement in the zero-degree direction. A "house" material will be determined during an initial benchmark specification project. This material will be readily and consistently available for the foreseeable future and will be processed by independent organizations using standard industry practice. These materials will be subsequently available for future simulation of the benchmark.

The only information necessary for the benchmark to be commenced by a participant is the external geometry information, the Phase 1 performance specification, and the "house" material specification and test results based on the future standard development procedures. The participant will use either the "house" material data or an alternative material of their choice providing they provide access to the raw material for processing (both test pieces and component manufacture) and the materials properties that are used in the simulation and development of the benchmark solution. The participant will use their analysis capabilities to develop the composite design for the provided specified external geometry to achieve the performance objectives specified. The initial benchmark submission will be supplemented by a manufacturing lay-up and specification suitable for a nominated manufacturer to develop a structure.

The benchmark coordinator will procure a sample manufactured to the participants' specification in either the house or the provided designated material. The item will be tested against the original requirement specification and if the item performs as predicted then the participant will be invited to analyze another load case on the same model. This will typically be at a completely different direction from the original specification. The participant will be invited to predict how it will perform and the coordinator will procure an identical component from the same manufacturer and test it accordingly. The emphasis on the second test is on the ability to predict failures in the non-preferred impact direction. It is expected that significantly less crush will occur and a premature catastrophic failure may be evident.

This approach to a universal benchmark has advantages when the potential participants have a disparate level of capability and experience. It allows all parties to investigate the true level of predictive capability for a a crush-dominated failure was well as a premature failure. It also provides an element of competition for both material suppliers and the participants themselves: the benchmark can be conducted with cut-off dates, where the participants can compete to specify the lightest structure using the "house" material in the first round. In the future as more materials become characterized they too can be offered as alternative materials and the benchmarks repeated.

5.4.2.5 Revival of the ACC Focal Project 3 Whole Vehicle Crash Analysis

The final recommendation is that the DOE/USCAR ACC Focal Project 3 whole vehicle crash analysis effort be revived. This suggestion was received following the August 2008 panel of experts meeting [2] and is reported by Brecher et at. [3]. Considerable research efforts were focused towards the development of the composites intensive Body In White (BIW). Although accepted at the outset that crash analysis capability was not available during the project, best practices from the racing car industry were employed in order to give the fundamental design a good prospect of stable crush without premature catastrophic collapse of the safety cell [62].

The design of the BIW already exists and the materials and thicknesses are defined to achieve the durability and static performance while permitting a 67% weight reduction over the conventional steel BIW. However, it may be appropriate to change the material type and processing method for more cost-effective prototyping while maintaining comparable performance. It is recommended that using available crash analysis methodologies, full vehicle analysis be performed to analyze the structure in a front and side impact. Following preliminary analysis of the baseline (currently designed) structure, further iterations may be required to develop the performance to be in line with expectations for current safety regulations.

Following successful prediction of the vehicle crash performance, the building block approach should be applied to verify the performance at the component and sub-assembly levels. This exercise will provide further confidence in the predictions and possibly provide the impetus to prototype the BIW and perform impact tests for comparison with the analysis predictions.

5.5 Summary

Despite the potential for utilizing composites in the automotive industry, databases for the types of composite materials applicable to this industry lags behind those in use by the aerospace industry. Additionally, there are significant needs for a material database that includes specialized crashworthiness properties of automotive-grade composite materials. Currently, however, no standardized test methods exist for assessing the crashworthiness of composites. Thus, the development of a database that focuses on specialized crashworthiness properties of composites requires the development of standardized crashworthiness test methods.

Significant progress has been made in recent years towards the development of crashworthiness test methods for composite materials. Further development of a flat-coupon test method as well as a element-level tube test method is recommended such that both types of tests may be standardized for use in assessing the crashworthiness of composite laminates. Following the development of these test methods, crashworthiness testing of automotive composite is recommended for purposes of both material screening as well as to support model development efforts.

Although considerable progress is being made towards the development of modeling approaches for predicting the crush behavior of composite structures, additional research is recommended. In coordination with the recommended crashworthiness testing, it is recommended that research be focused on creating a suite of benchmark model validations supported with detailed experiments for use in further developing the predictive capabilities of proposed modeling approaches.

6. REFERENCES

1. "Senate Report 109-293 – Transportation, Treasury, Housing and Urban Development, the Judiciary, and Related Agencies Appropriations Bill, 2007," *The Library of Congress*, http://thomas.loc.gov/cgi-bin/cpquery/?&sid=cp109xbGcm&refer=&r_n=sr293.109&db_id=109&item=&sel=TOC_289875&
2. "The Safety Characterization of Future Plastic and Composite Intensive Vehicles (PCIVs)," Workshop for Subject Matter Experts, DOT/RITA Volpe Center, U.S. Department of Transportation, NHTSA, Cambridge, MA, August 4, 2008.
3. Brecher, A., Brewer, J., Summers, S., and Patel, S., "Characterizing and Enhancing the Safety of Future Plastic and Composite Intensive Vehicles (PCIVs)." *21st International Conference on the Enhanced Safety of Vehicles*, Stuttgart, Germany, June 15-17, 2009. http://www-nrd.nhtsa.dot.gov/pdf/esv/esv21/09-0316.pdf.
4. Brecher, A., "A Safety Roadmap for Future Plastics and Composites Intensive Vehicles," Report number DOT HS 810 863, Volpe National Transportation Systems Center, November, 2007, http://permanent.access.gpo.gov/lps97921/4680pciv_safetyroadmap-nov2007.pdf
5. "Plastic and Composite Intensive Vehicles: An Innovation Platform for Achieving National Priorities," *American Chemistry Council Automotive Learning Center*, http://www.plastics-car.com/s_plasticscar/bin.asp?CID=425&DID=1956&DOC=FILE.PDF, September, 2009.
6. Cheah, L., Evans, C., Bandivadekar, A., Heywood, J. B., "Factor of Two: Halving the Fuel Consumption of New U.S. Automobiles by 2035", MIT Laboratory for Energy and the Environment, Cambridge, Massachusetts, 2007. http://web.mit.edu/sloan-auto-lab/research/beforeh2/files/cheah_factorTwo.pdf
7. "Honda Fireblade, The History," http://www.honda-fireblades.co.uk/history.html
8. Brylawski, M. M. and Lovins, A. B., "Ultralight-Hybrid Vehicle Design: Overcoming the Barriers to Using Advanced Composites in the Automotive Industry," proceedings of the *1996 International SAMPE Symposium and Exhibition*, pp. 1432-1446. Available at http://old.rmi.org/images/other/Trans/T95-39_UHVDAdvCompAuto.pdf
9. Dang, J. N., "Preliminary Results Analyzing the Effectiveness of Electronic Stability Control (ESC) Systems," NHTSA Evaluation Note DOT HS 809 790, September 2004, http://www.nhtsa.gov/cars/rules/regrev/evaluate/809790.html
10. Consumers Union, Yonkers, NY, 10703, http://www.consumersunion.org/
11. National Highway Traffic Safety Administration, www.nhtsa.gov
12. Insurance Institute for Highway Safety, www.iihs.org
13. "Vehicle Safety: Opportunities Exist to Enhance NHTSA's New Care Assessment Program", U.S. Government Accountability Office Report GAO-05-370, April 2005, www.gao.gov/products/GAO-05-370
14. Thornton, P. H., "Energy Absorption in Composite Structures", *Journal of Composite Materials*, Vol. 13, No. 3, pp 247-262, 1979.
15. Herrmann, H. G., Mohrdieck, C., and Bjekovic, R., "Materials for the Automotive Lightweight Design." DaimlerChrysler Research & Technology (Research Center Ulm)

REFERENCES

      presentation to *FKA/IKA Conferences New Advances in Body Engineering*, Aachen, Germany, November 28, 2002, p 17.

16. Hamada, H., and Ramakrishna, S., Comparison of Static and Impact Energy of Carbon Fiber/PEEK Composites Tubes," *Composite Materials: Testing and Design (Twelfth Volume), ASTM STP 1274*, R. B. Deo and C.R. Saff, Eds., American Society for Testing and Materials, pp. 182-196, 1996.
17. "HexWeb® Honeycomb Energy Absorption Systems Design Data," Hexcel Corporation, March, 2005, http://www.hexcel.com/NR/rdonlyres/96FE250C-7BB1-4295-82C4-461A31CC97A0/0/HexWebHoneycombEnergyAbsorptionBrochure.pdf
18. Fatality Analysis Reporting System (FARS) Traffic Safety Facts 2008 Data Overview, National Highway Traffic Safety Administration, http://www-nrd.nhtsa.dot.gov/Pubs/811162.pdf
19. Insurance Institute for Highway Safety, "50 Years of Research and Communications," http://www.iihs.org/50th/default.html.
20. "Aluminium in Cars," European Aluminium Association, http://www.eaa.net/upl/4/en/doc/Aluminium_in_cars_Sept2008.pdf
21. Krebs, R., "New Auto Safety Research Evaluates Impact of Size and Weight on Accidents - Lighter /Large Vehicles Can Decrease Fatalities, Contrary to Common Belief," http://www.americanchemistry.com/s_plastics/doc.asp?CID=1080&DID=6594
22. Marino, A. M., "Regulation of Performance Standards Versus Equipment Specification with Asymmetric Information," *Journal of Regulatory Economic,* Vol. 14, No X, pp 5-18, 1998.
23. Federal Motor Vehicle Safety Standard FMVSS 208, "Occupant Crash Protection," *Code of Federal Regulations, Title 49 – Transportation, Part 571*, National Highway Traffic Safety Administration, U.S. Department of Transportation.
24. Federal Motor Vehicle Safety Standard FMVSS 214, "Side Impact Protection," *Code of Federal Regulations, Title 49 – Transportation, Part 571*, National Highway Traffic Safety Administration, U.S. Department of Transportation.
25. Federal Motor Vehicle Safety Standard FMVSS 216, "Roof Crush Resistance," *Code of Federal Regulations, Title 49 – Transportation, Part 571*, National Highway Traffic Safety Administration, U.S. Department of Transportation.
26. PMC Crashworthiness Group, Composite Materials Handbook CMH-17, http://www.cmh17.org/pmc/crashworthiness.aspx
27. "Building Block Approach for Composite Structures," Volume 3, Chapter 4, Composite Materials Handbook, CMH-17, Revision G. www.CMH17.org
28. C-Zone, Engenuity Limited, www.compositesanalysis.com/crash/simulation/CZone
29. DeVries, K.L. and Adams, D.O. "Mechanical Testing of Adhesive Joints," *Adhesion Science and Engineering – 1: The Mechanics of Adhesion*, D. A Dillard and A. V. Pocius, Ed., Elsevier, 2002.
30. "History of Automotive Composites," Automotive Composites Alliance, http://www.autocomposites.org/composites101/history.cfm
31. Savage, G., "Formula 1 Composites Engineering," *Engineering Failure Analysis*, Vol. 17, No. 1, pp 92-115, 2010.
32. "McLaren MP4-1", F1Technical, http://www.f1technical.net/f1db/cars/476
33. "Static Load Testing," Article 18, Formula 1 Regulations, Federation Internationale de l' Automobile,

http://www.formula1.com/inside_f1/rules_and_regulations/technical_regulations/8707/fia.html
34. Savage, G. and Oxley, M., "Damage Evaluation and Repair of Composite Structures," *Anales de Mecánica de la Fractura,* Vol. 25, No. 2, 2008.
35. "Mulsanne's Corner: 1999 Mercedes-Benz CLR," http://www.mulsannescorner.com/benzCLR1.html
36. "Le Mans 99' Mercedes CLR-GT1 Crash Live," http://www.youtube.com/watch?v=Ow3rxq7U1mA&feature=related
37. "When Le Mans Racecars Fly", http://www.popsci.com/cars/article/2008-06/when-le-mans-racecars-fly
38. http://img91.imageshack.us/i/guest600yv1qi2.jpg/
39. "Mulsanne's Corner: Peter Elleray on the Bentley LMGTP," http://www.mulsannescorner.com/bentleyelleray.html
40. "McLaren F1 Crash Test," http://www.youtube.com/watch?v=mUPq760LC00
41. "2005 Porsche Carrera GT," http://www.wreckedexotics.com/carreragt/carreragt_20080911_002.shtml
42. "2005 Porsche Carrera GT," http://www.wreckedexotics.com/carreragt/carreragt_20061128_005.shtml
43. "2005 Porsche Carrera GT," http://www.wreckedexotics.com/carreragt/carreragt_20061128_003.shtml
44. "2005 Porsche Carrera GT," http://www.wreckedexotics.com/carreragt/carreragt_20051024_001.shtml
45. "2005 Porsche Carrera GT," http://www.wreckedexotics.com/carreragt/carreragt_20050602_002.shtml
46. "2005 Porsche Carrera GT," http://www.wreckedexotics.com/carreragt/carreragt_20050115_003.shtml
47. "Bugatti EB110," http://www.wreckedexotics.com/eb110/bugatti_20060328_001.shtml
48. "2003 Ferrari Enzo," http://www.wreckedexotics.com/enzo/enzo_20061205_004.shtml
49. "2003 Ferrari Enzo," http://www.wreckedexotics.com/enzo/enzo_20061205_002.shtml
50. "2003 Ferrari Enzo," http://www.wreckedexotics.com/enzo/enzo_20070428_001.shtml
51. "2003 Ferrari Enzo," http://www.wreckedexotics.com/enzo/enzo_20040205_005.shtml
52. "2003 Ferrari Enzo," http://www.wreckedexotics.com/enzo/enzo_20060809_007.shtml
53. "2003 Ferrari Enzo," http://www.wreckedexotics.com/enzo/enzo_20060809_006.shtml
54. "2005 Mercedes-Benz McLaren SLR," http://www.wreckedexotics.com/slr/slr_20080716_103.shtml
55. "2005 Mercedes McLaren SLR," http://www.wreckedexotics.com/slr/slr_20050504_001.shtml
56. "2004 Ferrari Enzo," http://www.wreckedexotics.com/enzo/enzo_20051103_004.shtml
57. "Porsche 911 Turbo," http://www.wreckedexotics.com/newphotos/bestof2010apr22/911_20090224_001.shtml
58. "Ferrari 360 Modena," http://www.wreckedexotics.com/newphotos/bestof2010apr22/360_20090311_001.shtml
59. "Porsche Boxter," http://www.wreckedexotics.com/boxster/boxster_20040414_003.shtml
60. United States Council for Automotive Research, LLC, http://www.uscar.org/guest/index.php

61. "Automotive Composites Consortium", United States Council for Automotive Research, LLC, http://www.uscar.org/guest/view_team.php?teams_id=25
62. Boeman, R. G. and Johnson, N. L., "Development of a Cost Competitive, Composite Intensive, Body- In-White," SAE Paper No. 2002-01=1905, *SAE Future Car Congress*, Crystal City, VA, June 2002.
63. Shaw, J. and Dodworth, A., "Metal With a Dark Side", keynote presentation, *22nd American Society for Composites Technical Conference*, Seattle, WA, September, 2007.
64. Verpoest, I., Thanh, T. C., and Lomov, S., "The TECABS Project: Development of Manufacturing, Simulation and Design Technologies for a Carbon Fiber Composite Car," proceedings of the *9th Japan International SAMPE Symposium*, pp 56-61, 2005.
65. "EU Transport Research – Success for Low-Weight Auto Parts Project," http://ec.europa.eu/research/transport/news/article_1507_en.html
66. Greve, L. and Pickett, A., "Modelling Damage and Failure in Carbon/Epoxy Non Crimp Fabric Composites Including Effects of Fabric Pre-Shear," *Composite Part A: Applied Science and Manufacturing*, Vol. 37, No. 11, pp 1983-2001, 2006.
67. Goede, M., Stehlin, M., Rafflenbeul, L., Kopp, G., and Beeh, E., "Super Light Car— Lightweight Construction Thanks to a Multi-Material Design and Function Integration," *European Transport Research Review*, Vol. 1, No. 1, pp 5-10, 2008.
68. Taketa, I., Yamaguchi, K., Wadahara, E., Yamasaki, M., Sekido, T., and Kitano, A., "The CFRP Automobile Body Project in Japan," Proceedings of the *Twelfth US-Japan Conference on Composite Materials*, University of Michigan-Dearborn, September 21-22, 2006.
69. Sills, P., "A Novel Solution for Achieving Lightweight, Safe Vehicle Structures Through Composites," *Symposium on Reducing Greenhouse Gas Emissions from Vehicles*, Sacramento, CA, April 21, 2008.
70. Warren, D., ORNL Carbon Fiber Technology Development, Oak Ridge National Laboratories, Oak Ridge, Tennessee.
71. "New Carbon Fiber Plant to be Built in Moses Lake, WA", SGL Group and BMW Group Press Release, April 6, 2010.
72. "Frontal Offset Crashworthiness Evaluation", *2002 Insurance Institute for Highway Safety Guidelines for Rating Structural Performance*.
73. Roberts, R. and Barnes, G, "Flat Coupon Testing Considerations," *54th CMH-17 Meeting*, Crashworthiness Working Group, Salt Lake City, UT, March 2009.
74. Dean, G., and Crocker, L., "A Proposed Failure Criterion for Tough Adhesives," Project PAJ2 NPL Report No.10 CMMT(a), February, 1999.
75. Chung, J., Chandhery, K., "Roles of Discontinuities in Bio Inspired Adhesive Pads," *Journal of the Royal Society Interface*. Vol. 2, No. 2, pp 55–61, March 2005.
76. Conrad, M., Smith, G. and Fernand, G., "Fracture of Discontinuous Wood Adhesive Bonds," *International Journal of Adhesion and Adhesives*, Vol. 23, No. 1, pp. 39-47.
77. Brewer, J., US Patent Application, 08/546,388, October 1995.
78. Barnes, G., Coles, I., Johnson, N., and Boeman, R., "ACC Focal Project 3: Analysis and Design of Composite Intensive BIW", ICCM 14, San Diego, CA, July 14-18, 2003.
79. Barnes, G., Engenuity Limited, www.compositesanalysis.com/crash/simulation/casestudy/PlainCone
80. Roberts, T., "The Carbon Fiber Industry Worldwide 2008-2014", Tony Roberts, ISBN 1871677599, updated June 2009.

REFERENCES

81. "National Institute for Aviation Research: NCAMP Overview", http://www.niar.wichita.edu/coe/ncamp.asp
82. Xiao, X., "Simulation of Composite Tubes Axial Impact with a Damage Mechanics Based Composite Material Model," *10th International LS-DYNA Users Conference*, Dearborn, MI, June 8-10, 2008.
83. Barnes, G., "CZone Composite Crush Predictions", *49th MIL-HDBK17 Meeting*, Santa Monica, CA, December, 2005.
84. Czaplicki, M.J., Robertson, R.E., Thornton, P.H., "Comparison of Bevel and Tulip Triggered Pultruded Tubes for Energy Absorption", *Composites Science and Technology*, Vol. 40, No. 1, pp 31-46, 1990.
85. Thornton, P.H., "Effect of Trigger Geometry on Energy Absorption in Composite Tubes," *Fifth International Conference on Composite Materials*, Metallurgical Society Inc., pp 1183-1199, 1985.
86. Hanagud, S., Craig, J.I., Sriram, P., and Zhou, W., "Energy Absorption Behavior of Graphite Epoxy Composite Sine Webs", *Journal of Composite Materials*, Vol. 23, No. 5, pp 448-459, 1989.
87. Lavoie, A. and Kellas, S., "Dynamic Crush Tests of Energy-Absorbing Laminated Composite Plates," *Composite - Part A: Applied Science and Manufacturing*, Vol. 27, No. 6, pp. 467-475, 1996.
88. Bolukbasi, A.O. and Laananen, D.H., "Energy Absorption in Composite Stiffeners," *Composites*, Vol. 26, No. 4, pp 291-301, 1995.
89. Fairfull, A.H and Hull, D., "Energy Absorption of Polymer Matrix Composite Structures: Frictional Effects", in *Structural Failure*, ed. T. Wierzbicki, and N. Jones, John Wiley & Sons, Inc., Chapter 8, pp 255-279, 1989.
90. Feraboli, P., "Development of a Corrugated Test Specimen for Composite Materials Energy Absorption," *Journal of Composite Materials*, Vol. 42, No. 3, pp 229-256, 2008.
91. Garner, D. and Adams, D. O., "Flat Coupon Test Method for Composite Crashworthiness", *54th CMH-17 Meeting*, Crashworthiness Working Group, Salt Lake City, UT, March 2009.
92. Warrior, N., Turner, T., Cooper, T. and Ribeaux, M., Effects of Boundary Conditions on the Energy Absorption of Thin-Walled Polymer Composite Tubes under Axial Crushing, *Thin-Walled Structures*, Vol. 46, No. 7-9, July-September 2008.
93. Brimhall, T. J., "Friction Energy Absorption in Fiber Reinforced Composites," Ph.D. Dissertation, Department of Mechanical Engineering, Michigan State University, 2005.
94. Stapleton, S.E. and Adams, D.O., "Crush Initiators for Increased Energy Absorption in Sandwich Composites," *Journal of Sandwich Structures and Materials*, Vol. 10, No. 4, pp 331-354, 2008.
95. A. Johnson, Determination of Composite Energy Absorption Properties, *Proceedings of the 50th CMH-17 Coordination Group Meeting – Polymer Matrix Composite Materials Handbook Meeting*, pp 477-531, 2006.
96. Zhou, W., Craig, J.I., and Hanagud, S.V., "Crashworthy Behavior of Graphite/Epoxy Composite Sine Wave Webs," *Proceedings of the 32nd AIAA/ASME/ASCE/AHS/ASC Structures, Structural Dynamics, and Materials Conference*, pp. 1618-1626, 1991.
97. Lavoie, J.A. and Morton, J., "A Crush Test Fixture for Investigating Energy Absorption of Flat Composite Plates," *Experimental Techniques*, Vol. 18, No. 6, pp 23-26, 1994.

98. Lavoie, J.A., Morton, J., Kellas, S., and Jackson, K., "New Test Method for Measuring Static and Dynamic Energy Absorption Capacity of Composite Plates, *Advancing with Composites*, pp 239-250, 1994.
99. Lavoie, J.A., Morton, J., and Jackson, K., "An Evaluation of the Energy Absorption of Laminated Composite Plates," *Proceedings of the Institution of Mechanical Engineers, Part G: Journal of Aerospace Engineering*, Vol. 209, No. 3, pp 185-194, 1995.
100. Bolukbasi, A.O. and Laananen, D.H., "Energy Absorption in Composite Stiffeners", *Composites*, Vol. 26, No. 4, pp 291-301, 1995.
101. Dubey, D.D. and Vizzini, A.J., "Testing Methods for Energy Absorption of Kevlar/Epoxy," *Journal of the American Helicopter Society*, Vol. 44, No. 3, pp 179-187, 1999.
102. Cauchi Savona, S., and Hogg, P.J. "Investigation of Plate Geometry on the Crushing of Flat Composite Plates," *Composites Science and Technology*, Vol. 66, No. 11-12, pp 1639-1650, 2005.
103. Engenuity Limited, www.compositesanalysis.com/crash/testing/dynamic/crushfixtures
104. Takashima, T., Ueda, M., and Kato, Y., "Experimental Study on Energy Absorption of CFRP Laminated Plate Using New Test Fixture," *10th Japan International SAMPE Symposium & Exhibition*, 2007.
105. Feraboli, P., "Development of a Modified Flat-plate Test Specimen and Fixture for Composite Materials Crush Energy Absorption," *Journal of Composite Materials*, Vol. 43, No. 19, pp 1967 – 1990, 2009.
106. Mamalis, A.G, Manolakos, D.E., Papapostolou, D.P., and Ioannidis, M.B., "On The Response Of Thin-Walled CFRP Composite Tubular Components Subjected To Static and Dynamic Axial Compressive Loads," *Composite Structures*, Vol. 69, No. 4, pp 407-420, 2005.
107. Melo, J.D., Silva, A.L.S., Villena, J.E., "The Effect Of Processing Conditions On The Energy Absorption Capability Of Composite Tubes," *Composite Structures*, Vol. 82, No. 4, pp 622-628, 2008.
108. Yang, Y., Nakai, A., Uozumi, T., Hamada, H., "Energy Absorption Capability of 3D Braided Textile Composite Tubes With Rectangular Cross Section," *Key Engineering Materials*, Vol. 334-335, pp 581-584, 2007.
109. Hull, D., "A Unified Approach to Progressive Crushing of Fiber-Reinforced Composite Tubes," *Composite Science and Technology*, Vol. 40, No. 4, pp 377-421, 1991.
110. Caruthers, J.J., Kettle, A.P, and Robinson, A.M., "Energy Absorption Capability and Crashworthiness of Composite Material Structures: A Review," *Applied Mechanics Review*, Vol. 51, No. 10, pp 635-649, 1998.
111. Jacob, G.C., Fellers, J.F., Simunovic, S., and Starbuck, J.M., "Energy Absorption in Polymer Composites for Automotive Crashworthiness," *Journal of Composite Materials*, Vol. 36, No. 7, pp 813-850, 2002.
112. Elgalai, A.M., Hamouda, A.M.S., and Sahari, B.S., "Energy Absorption Capabilities of Woven Roving Glass/Epoxy Composite Tubes: Effect of Tube Length," *Strength, Fracture, and Complexity*, Vol. 3, pp 15-24, 2005.
113. Mamalis, A.G., Manolakos, D.E., Demosthenous, G.A., and Ioannidis, M.B., "Crashworthiness of Composite Thin-Walled Structural Components," Technomic Publishing Co., Inc., Lancaster, 1998.

REFERENCES

114. Bisagni, C., Di Peitro, G., Fraschini, L., and Terletti, D., "Progressive Crushing of Fiber-Reinforced Composite Structural Components of a Formula One Racing Car," *Composite Structures*, Vol. 68, No. 4, pp. 491-503, 2004.
115. Johnson, N.L., Browne, A.L., Watling, P.J., and Peterson, D.G., "Parameter Effects on the Dynamic Crush Performance of Braided "Hourglass" Cross Section Composite Tubes," *Proceedings of the 9th Annual ASM/ESD Advanced Composites Conference, Advanced Composites Technologies*, pp 403-419, 1993.
116. Jacob, G.C., Fellers, J.F., Starbuck, J.M., and Simunovic, S., "Crashworthiness of Automotive Composite Material Systems," *Journal of Applied Polymer Science*, Vol. 92, No. 5, pp 3218-3225, 2004.
117. Jacob, G.C., Starbuck, J.M., Fellers, J.F., Simunovic, S., and Boeman, R.G., "Strain Rate Effects on the Mechanical Properties of Polymer Composite Materials," *Journal of Applied Polymer Science*, Vol. 94, No. 1, pp 296-301, 2004.
118. Jacob, G.C., Starbuck, J.M., Fellers, J.F., Simunovic, S., and Boeman, R.G., "The Effect of Loading Rate on the Fracture Toughness of Fiber Reinforced Polymer Composites," *Journal of Applied Polymer Science*, Vol. 96, No. 3, pp 899-904, 2005.
119. Jacob, G.C., Starbuck, J.M., Fellers, J.F., Simunovic, S., and Boeman, R.G., "Crashworthiness of Various Random Chopped Carbon Fiber Reinforced Epoxy Composite Materials and Their Strain Rate Dependence," *Journal of Applied Polymer Science*, Vol. 101, No. 3, pp 1477-1486, 2006.
120. Farley, G.L. and Jones, R.M., "Crushing Characteristics of Continuous Fiber-Reinforced Composite Tubes," *Journal of Composite Materials*, Vol. 26, No. 1, pp 37-50, 1992.
121. Bruce, D.M, Matlock, D.K., Speer, J.G., De, A., "Assessment of the Strain-Rate Dependent Tensile Properties of Automotive Sheet Steels," *SAE 2004 World Congress*, Doc. No. 2004-01-0507, Detroit, MI, March 2004.
122. Smerd, R., Winkler, S., Salisbury, C., Worswick, M., Lloyd, D., and Finn, M. "High Strain Rate Tensile Testing of Automotive Aluminum Alloy Sheet," *International Journal of Impact Engineering*, Vol. 32, No. 4, pp 541-560, 2005.
123. "Recommendations for Dynamic Tensile Testing of Sheet Steels," International Iron and Steel Institute (IISI), August 2005, http://www.worldautosteel.org/uploaded/DynTestingRecomPract.pdf
124. Wood, P.K.C. and Schley, C.A., *"Strain Rate Testing of Metallic Materials and their Modeling for use in CAE based Automotive Crash Simulation Tools (Recommendations & Procedures),"* Smithers Rapra Technology, 2009.
125. "Plastics - Determination of Tensile Properties at High Strain Rates," ISO 18872:2007, International Organization for Standardization, 2007.
126. "High Strain Rate Tensile Testing of Polymers," SAE J2749-2008, SAE International, 2008.
127. Simón, J.C., Johnson, E., and Dillard, D.A., "Characterizing Dynamic Fracture Behavior of Adhesive Joints under Quasi-Static and Impact Loading," *Journal of ASTM International*, Vol. 2, No. 7, July/August, 2005.
128. Pohlit, D. J., Dillard, D.A., Jacob, G.C., and Starbuck, J.M., "Evaluating the Rate-Dependent Fracture Toughness of an Automotive Adhesive," Journal of Adhesion, Vol. 84, No. 2, pp 143-163, 2008.
129. Hampton AutoBeat LLC, "The Daily Update on Automotive Technologies," *Autotech Daily*, 2002.

130. Gohlami, T., Lescheticky J., and Paßmann, R., "Crashworthiness Simulation of Automobiles with ABAQUS/Explicit", 2003 ABAQUS Users' Conference.
131. Spethmann, P. and Hersatt, C. "Crash Simulation Evolution and its Impact on R&D in Automotive Applications," *International Journal Product Development*, Vol. 8, No.3, 2009.
132. Siljander, A., A Review of Aeronautical Fatigue Investigations in Finland During the Period April 2003 to April 2005, Presented at the 29th Conference of the International Committee on Aeronautical Fatigue (ICAF), Hamburg, Germany, June 2005.
133. Hallquist, J., "LS-Dyna Theoretical Manual," Livermore Software Technology Corporation, Livermore, CA, 2006 http://www.lstc.com/pdf/ls-dyna_theory_manual_2006.pdf
134. Pinho, S., Iannucci, L. and Robinson, P., "Physically-Based Failure Models and Criteria for Laminated Fiber-Reinforced Composites. Part II: FE Implementation," *Composites Part A: Applied Science and Manufacturing*, Vol. 37, No. 5, pp 766-777, 2006.
135. Chang, F. K. and Chang, K. Y., "A Progressive Damage Model for Laminated Composites Containing Stress Concentrations," *Journal of Composite Materials*, Vol. 21, No. 9, pp. 834-855, (1987).
136. Tsai, S. W. and Wu, E. M., "A General Theory of Strength for Anisotropic Materials," *Journal of Composite Materials*, Vol. 5, pp. 58-80, (1971).
137. Feraboli, P., Standardization of Numerical and Experimental Methods for Crashworthiness Energy Absorption of Composite Materials, JAMS Annual Meeting, June 2008, Everett, WA
138. "ABAQUS/Explicit", Simulia, Providence, RI, 2009, www.simulia.com/products/abaqus_explicit.html
139. "RADIOSS: A Complete Finite Element Solver for Structural Analysis," Altair Engineering, Inc., 2009, www.radioss.com/pdfs/product_brochures/HW_RADIOSS_Web.pdf
140. "PAM-CRASH: Virtual Performance for Crash & Safety Professionals," ESI Group, 2009, www.esi-group.com/products/crash-impact-safety/pam-crash
141. Feraboli, P., and Rassaian, M., "Standardization of Analytical and Experimental Methods for Crashworthiness Energy Absorption of Composite Materials," Presented at the *Federal Aviation Administration Joint Advanced Materials & Structures (JAMS) 5th Annual Technical Review Meeting*, Wichita, KS July 21-22, 2009.
142. Xiao, X. and Sheidaei, A., "Simulation of Multiple Shapes Specimens Using LS-DYNA MAT58 and User Material Model," *55th CMH-17 Meeting*, Crashworthiness Working Group, Atlanta, GA, November 18, 2009.
143. Matzenmiller, A., Lubliner, J., Taylor, R, A Constitutive Model for Anisotropic Damage In Fiber Composites, *Mechanics of Materials*, Vol. 20, No. 2, pp 125-152, 1995.
144. Gatti, M., Vescovi, L., Sperati, M., Pagano, P. and Ferrero, L., "Characterization of Composite Materials, Unidirectional and Fabric Samples," Autosim Technology Workshops & Csc, Paris, France, July 5-6, 2007.
145. Schweizerhof, K., Weimar, K., Munz, and Rottner, T.H., "Crashworthiness Analysis and Enhanced Composite Material Models in LS-DYNA – Merits and Limits," *LS-Dyna World Conference*, 1998.

146. Flesher, N.D. and Chang, F.K., "Dynamic Tube Crush Simulation with Explicit FEM," presentation to the Energy Management Working Group of the Automotive Composites Consortium, Southfield, Michigan, January, 2005.
147. Flesher, N.D., "Crash Energy Absorption of Braided Composite Tubes," Ph.D. dissertation, Department of Mechanical Engineering, Stanford University, 2006.
148. Fish, J. and Yuan, Z., "Multiscale Modeling for Crash Prediction of Composite Structures: 2008 Report," annual report submitted to the Energy Management Working Group of the Automotive Composites Consortium, January, 2009.
149. McGregor, C., Zobeiry, N., Vaziri, Reza, and Poursartip, A., "A Constitutive Model for Progressive Compressive Failure of Composites", *Journal of Composites Materials*, Vol. 42, No. 25, pp 2687-2716, 2008.
150. "CZone for ABAQUS", Simulia, Providence, RI, 2008, www.simulia.com/download/products/CZoneFlyer.pdf
151. US Crush Modeling Patent Application US Patent No. 7,630,871, Engenuity Limited, www.compositesanalysis.com/crash/simulation/IP/PatentNo7630871
152. Greve, L., Pickett, A. and Payen, F., "Experimental Testing and Phenomenological Modeling of the Fragmentation Process of Braided Carbon/Epoxy Composite Tubes Under Axial and Oblique Impact," *Composites Part B: Engineering*, Vol. 39, No. 7-8, pp 1221-1232, 2008.
153. Roberts, R., "Engenuity's Simulation of the Semi-Circular Corrugated Specimen", *53rd CMH-17 Meeting*, Crashworthiness Working Group, Ottawa, Canada, August 2008.
154. Indermuehle, K., "Composite Crashworthiness Modeling using ABAQUS/Explicit," *55th CMH-17 Meeting*, Crashworthiness Working Group, Atlanta, GA, November, 2009.
155. Dodworth, A., Barnes, G., and Roberts, R., "Composite Crushing Simulation Using CZone and Abaqus", Abaqus User Group, 2008.
156. Tsai, S. W., "Strength Theories of Filamentary Structures," in R. T. Schwartz and H. S. Schwartz (eds.), *Fundamental Aspects of Fiber Reinforced Plastic Composites*, pp 3-11, Chap. 1, Wiley Interscience, New York (1968).
157. MAT8 Post Failure Degradation Rules, MSC.Dytran User's Manual, Section 2-32, MSC Software Corporation, Santa Ana, CA.
158. Hashin, Z., "Failure Criteria for Unidirectional Fiber Composites," Journal of Applied Mechanics, Vol 47, No. 2, pp. 329-334, (1980).
159. Private communication with ABAQUS/CZone Code Developers.
160. Pinho, S., Robinson, P., and Iannucci, L. "Fracture Toughness of the Tensile and Compressive Fiber Failure Modes in Laminated Composites," *Composites Science and Technology*, Vol. 66, No. 13, pp 2069-2079.
161. Christensen, R., "Failure of Fiber Composite Laminates: Progressive Damage and Polynomial Invariants," 2008, www.failurecriteria.com
162. Knight, F. Jr., "User-Defined Material Model for Progressive Failure Analysis," NASA/CR-2006-214526, 2006.
163. "Roadmap for the Numerical Round Robin," *55th CMH-17 Meeting*, Crashworthiness Working Group, Atlanta, GA, November 2009.
164. Feraboli, P., "Standardization of Numerical and Experimental Methods for Composite Crashworthiness – Year II," Presented at the *Federal Aviation Administration AMTAS Fall Meeting*, Seattle, WA, November, 2008.

REFERENCES

165. Roberts, R., "Engenuity's Simulation Of The Round 1 and Round 2 Specimen," *55th CMH-17 Meeting*, Crashworthiness Working Group, Atlanta, GA, November, 2009.
166. Feraboli, P., 53rd CMH-17 Meeting, Crashworthiness Working Group, Ottawa, Canada, August 2008.
167. Johnson., A, David., M., "CMH-17 Crashworthiness WG Round Robin Simulation Of Crash Elements," *55th CMH-17 Meeting*, Crashworthiness Working Group, Atlanta, GA, November, 2009.
168. Foedinger, R, Meeker, J. "CMH-17 Crashworthiness Forum Round 2 - C-Channel Specimen LS-DYNA MAT162 Simulation," *55th CMH-17 Meeting*, Crashworthiness Working Group, Atlanta, GA, November 18, 2009.
169. "Crashworthiness Forum" *55th CMH-17 Meeting*, Crashworthiness Working Group, Atlanta, GA, November 18, 2009.
170. Deleo, F., Feraboli, P., and Rassaian, M., "Preliminary Results of Stage II of the Round Robin, MAT54," *55th CMH-17 Meeting*, Crashworthiness Working Group, Atlanta, GA, November 18, 2009.
171. Deleo, F., Feraboli, P., and Rassaian, M., "Preliminary Results of Stage II of the Round Robin, MAT58," *55th CMH-17 Meeting*, Crashworthiness Working Group, Atlanta, GA, November 18, 2009.
172. Caliskan, A., "Component Impact Analysis Using RADIOSS: Stage II Results," *55th CMH-17 Meeting*, Crashworthiness Working Group, Atlanta, GA, November 18, 2009.
173. Moulliet, J. B., "Atlanta: RRI part 2", *55th CMH-17 Meeting*, Crashworthiness Working Group, Atlanta, GA, November 18, 2009.
174. Barnes, G., "Engenuity's Simulation of the Round 1 and Round 2 Specimen," *55th CMH-17 Meeting*, Crashworthiness Working Group, Atlanta, GA, November 18, 2009.

www.ingramcontent.com/pod-product-compliance
Lightning Source LLC
Chambersburg PA
CBHW080256180526
45167CB00006B/2544